T0282444

Inspiring Conversations with Women Professors

Inspiring Conversations with Women Professors

The Many Routes to Career Success

Anna M. Garry

ETH Zürich, Zurich, Switzerland

Academic Press is an imprint of Elsevier
125 London Wall, London EC2Y 5AS, United Kingdom
525 B Street, Suite 1650, San Diego, CA 92101, United States
50 Hampshire Street, 5th Floor, Cambridge, MA 02139, United States
The Boulevard, Langford Lane, Kidlington, Oxford OX5 1GB, United Kingdom

Copyright © 2019 Elsevier Inc. All rights reserved.

No part of this publication may be reproduced or transmitted in any form or by any means, electronic or mechanical, including photocopying, recording, or any information storage and retrieval system, without permission in writing from the publisher. Details on how to seek permission, further information about the Publisher's permissions policies and our arrangements with organizations such as the Copyright Clearance Center and the Copyright Licensing Agency, can be found at our website: www.elsevier.com/permissions.

This book and the individual contributions contained in it are protected under copyright by the Publisher (other than as may be noted herein).

Notices
Knowledge and best practice in this field are constantly changing. As new research and experience broaden our understanding, changes in research methods, professional practices, or medical treatment may become necessary.

Practitioners and researchers must always rely on their own experience and knowledge in evaluating and using any information, methods, compounds, or experiments described herein. In using such information or methods they should be mindful of their own safety and the safety of others, including parties for whom they have a professional responsibility.

To the fullest extent of the law, neither the Publisher nor the authors, contributors, or editors, assume any liability for any injury and/or damage to persons or property as a matter of products liability, negligence or otherwise, or from any use or operation of any methods, products, instructions, or ideas contained in the material herein.

British Library Cataloguing-in-Publication Data
A catalogue record for this book is available from the British Library

Library of Congress Cataloging-in-Publication Data
A catalog record for this book is available from the Library of Congress

ISBN: 978-0-12-812346-1

For Information on all Academic Press publications
visit our website at https://www.elsevier.com/books-and-journals

Publisher: Mica Haley
Acquisition Editor: Mary Preap
Editorial Project Manager: Megan Ashdown
Production Project Manager: Poulouse Joseph
Cover Designer: MPS

Typeset by MPS Limited, Chennai, India

Contents

List of participants

Natalie Banerji Department of Chemistry and Biochemistry, University of Bern, Bern, Switzerland

Eleni Chatzi Department of Civil, Environmental and Geomatic Engineering, ETH Zürich, Zurich, Switzerland

Emanuela Del Gado Department of Physics, Georgetown University, Washington, DC, United States

Rachel Grange Department of Physics, ETH Zürich, Zurich, Switzerland

Stefanie Hellweg Department of Civil, Environmental and Geomatic Engineering, ETH Zürich, Zurich, Switzerland

Ursula Keller Department of Physics, ETH Zürich, Zurich, Switzerland

Salomé LeibundGut-Landmann VetSuisse Faculty, University of Zürich, Zurich, Switzerland

Ulrike Lohmann Department of Environmental Systems Science, ETH Zürich, Zurich, Switzerland

Marloes Maathuis Department of Mathematics, ETH Zürich, Zurich, Switzerland

Isabelle Mansuy Department of Health, Sciences and Technology, ETH Zürich, Switzerland; Department of Medicine, University of Zürich, Zurich, Switzerland

Paola Picotti Department of Biology, ETH Zürich, Zurich, Switzerland

Ursula Röthlisberger Institute of Chemical Sciences and Engineering, EPFL, Lausanne, Switzerland

Clara Saraceno Department of Electronics and Information Technology, Ruhr Bochum University, Bochum, Germany

Maria Schönbächler Department of Earth Sciences, ETH Zürich, Zurich, Switzerland

Olga Sorkine-Hornung Department of Computer Science, ETH Zürich, Zurich, Switzerland

Nicola Spaldin Department of Materials, ETH Zürich, Zurich, Switzerland

Elsbeth Stern Department of Humanities, Social and Political Sciences, ETH Zürich, Zurich, Switzerland

Shana Sturla Department of Health, Sciences and Technology, ETH Zürich, Switzerland

Effy Vayena Department of Health, Sciences and Technology, ETH Zürich, Switzerland

Brigitte von Rechenberg VetSuisse Faculty, University of Zürich, Zurich, Switzerland

Katharina von Salis (Retired professor since 1990s) Department of Earth Sciences, ETH Zürich, Zurich, Switzerland

Sabine Werner Department of Biology, ETH Zürich, Zurich, Switzerland

Marcy Zenobi-Wong Department of Health, Sciences and Technology, ETH Zürich, Switzerland

About the author

Anna Garry holds a BSc in Science and Technology Policy and a PhD in Political Science from the University of Manchester, United Kingdom. She has diploma in Adult Education Teaching (Manchester), and an MA in Creative Writing from the University of East Anglia (UEA). Her early research and training experience focused on developing and testing novel educational methods at the university level, in mainly STEM subjects, at Heriot-Watt University, Edinburgh (Civil Engineering); Moray House School of Education, Edinburgh University; the University of West of England, Bristol, and University of Manchester (Computer Science). The projects explored nontraditional methods of teaching such as experiential learning, creative and critical thinking techniques, the use of self-assessment in learning, and the effectiveness of on-line courses in teaching. She has designed and presented workshops on the continuing professional development skills for civil and structural engineers, and for PhDs and postdocs. She was a lecturer at UEA in the undergraduate creative writing program and, in parallel, she led, and expanded, the adult Continuing Education provision in Creative Writing.

Since 2010 she has worked as the Outreach Officer for the National Center for Competence in Research Molecular Ultrafast Science and Technology (NCCR MUST). She is responsible for the programs in advancement of women/equal opportunities, education and training, and scientific outreach.

Dr. Anna M. Garry
NCCR MUST, ETH Zürich, Zurich, Switzerland

Foreword

In 2010, when I was appointed the Director of the Swiss National Competence Center in Research, Molecular Ultrafast Science and Technology (NCCR MUST), as part of my remit, I was required by the Swiss National Science Foundation (SNSF) to undertake initiatives to advance the situation of women scientists in my discipline. This was an enormous opportunity for me to work to bring about significant changes for women scientists in our fields of physics and chemistry. The NCCR funding tool is, potentially, a 12-year funding instrument and the Swiss National Science Foundation assesses us annually on our achievements, not just in scientific research, but also under the outreach programs: equal opportunities, education and training, technology transfer, and communication. I always knew how important it is to have more visible role models in order to inspire young women to consider a career in science. From my perspective, it is equally important for excellent female academics to be more publicly visible in order to increase and deepen understanding of the significant research contributions of women professors in academia. This book project resulted from a number of initiatives in the first eight years of MUST, particularly the creation of a Women Professors Forum at ETH Zürich.

My experience of being the first tenured physics professor at ETH Zürich made me realize that, as a minority member of a department, one can feel pretty isolated in your local community. This experience gave me the idea, as director, to invest in the creation of a formal organization that would connect all the female professors at ETH Zürich, which is mainly a technical university. At this point, the percentage of tenured female professors at ETH stood at 8.5%, with the assistant professorship numbers standing at 28.5%. Given that there were few female professors across the university, it was clear that there could be real benefits to women professors to have a formal association that connected them through networking events and other activities. In the early days, we reached out to the Women Faculty Forum at Yale University to learn from the experience of an association, which was already 10 years old, established in 2001 during Yale's 300-year celebration of its foundation.[1] Then, working with a committee of senior female professors from ETH, we created an association, sponsored by NCCR MUST, that by the end of 2012 had already 80% membership of female professors in the university. It became known as the ETH Women Professors Forum (ETH WPF) and has now been in operation for six years.

Once established, the ETH Women Professors Forum launched a lunch-time meeting program where, initially senior professors, presented their research, and created a lunch exchange and discussion on career matters between members from

[1] Yale Women Faculty Forum, https://wff.yale.edu.

all stages of the career. The aims of the ETH WPF are to increase networking across the university, to enable members to have informal access to many different women professors for quick advice and exchange of experience—sometimes through a simple phone call—to provide visible role models to students and the community, and to develop political influence in order to improve the recruitment, advancement, and retention of female professors.[2] I was elected to be the President of an eight-person WPF Executive Board, and served two terms, from 2012 to 2016.

In 2016 under the newly elected chair of Prof. Janet Hering (Director of Eawag, the ETH Aquatic Research Institute and professor at EPFL), the association expanded to include the professors working at our sister university EPFL Lausanne, and the other research institutes connected to ETH Zürich. With this expanded membership, the participation of female professors built to 80% across the new participating institutions. The aims of supporting the community of women professors remain very much alive. The latest ETH gender statistics show that in 2017, the percentage of female full professors stands at 12%, while APs is at 22%, meaning that progress is slow and work needs to be done.

I am aware of just how much the existence of a Women Professors Forum could have helped me earlier in my university career, to integrate well into the university. The Forum acts as a sounding board and support structure for other women professors with similar experiences. I also learned a great deal, and grew as a leader, from working directly with other colleagues at the executive leadership level, such as Prof. Janet Hering as the director of Eawag.

It was a pleasure to take part in this project, by contributing my story to the conversations, but also by supporting Anna Garry's investment to produce this book. It is fascinating to see the range of different experiences of these women professors who are working in a wide selection of academic disciplines, and to see how they have developed their careers in contrasting ways. From my perspective, you have to follow your passion, be the best you can be and keep trying to understand who you are. Every person is different in the wishes and routes chosen, but if you are following your instincts, aiming for excellence at all levels, learning to trust who you are, accepting that you cannot succeed in everything, developing your frustration tolerance, and simply try again if it is important for you: you will succeed in what you do, whatever choices you make in life.

I wish future generations much joy and success in following paths that we, in this book, have taken!

Prof. Ursula Keller
Department of Physics, ETH Zürich, Zürich, Switzerland

[2] An interview with Ursula Keller by Andrea Eichholzer in *Leadership in Universities* (Führen in Hochschulen), Springer, 2017, "*Development of a network for women in academic leadership positions: What lessons can be learned from the ETH Women Professors Forum?*" http://www.nccr-must.ch/nccr_must/news_4.html?3923.

Acknowledgments

I would like to thank Prof. Ursula Keller, Director of NCCR MUST, for her ongoing and significant commitment to gender equality issues, her dedication to making the expertise of successful women professors visible, and for the support she has given, which enabled me to invest in this book project in the last years.

Thanks also to Prof. Thomas Feurer, NCCR MUST Co-Director for his dedication to young career researchers and his commitment to our equal opportunity program.

The Swiss National Science Foundation finances NCCR MUST and I would like to thank them for their sponsorship.

In Switzerland, there is a strong community dedicated to equal opportunities. Many colleagues have influenced and motivated me considerably; thanks to Maya Widmer, Monika Keller, Gabriela Obexer-Ruff, Irene Rehmann, Simona Isler, Karin Gilland-Lutz, Christiane Löwe, Elisabeth Maurer, Mihaela Falub, Kristin Hoffmann, and Honorata Kaczykowski-Patermann.

I spoke, informally, with a large number of female professors at events, conferences, and scientific meetings over the last four years and would like to thank them for their openness about their lives, careers, and research expertise.

I would like to thank Prof. Silvia Dorn (Emeritus professor, ETH Zürich) for her insight, extensive experience and generous support during the process of bringing this book to fruition.

Thanks also to my colleagues Jan van Beilen and Nadia Sigrist for the great working environment.

To my good friends Jane Williams and Tanja Grobelnik: thank you for your humor, warmth and ongoing support over the last few years.

Introduction

Origins and influences

The starting point of this book was to gather the stories and experiences of exceptional women in academia, stories about women from numerous nationalities working in university environments in Switzerland. They remain a minority, and I wanted to make visible their research work and to celebrate their achievements. Through working also with aspiring young researchers, particularly young women, I wanted also to provide them with valuable information and insight into the various routes to professorship arising from the career experience of female professors. This book project has evolved and developed over the last four years.

Three initiatives had a strong influence on the book, all of them originated from the nationwide scientific network, the NCCR MUST: the establishment of the ETH Women Professors Forum in 2012; the ongoing work to support the career development and goals of young researchers in the MUST network; and the establishment of a women scientist network for PhDs and postdocs in 2011.

The first key experience was the creation of the Women Professors Forum at ETH, upon Ursula Keller's initiative, with its aim to interconnect the relatively few women professors working at a mainly scientific and technical university. Through supporting the establishment of this network, I had the opportunity to meet many female professors across the university's departments who were active at different stages of a professorial career. The first consultation process, which was aimed at identifying the wishes of women professors for the planned network, helped me to gain some insight into the career at different stages. As we established the Women Professor Forum networking lunches, where senior female professors highlighted their research, opened their laboratories to their colleagues, and shared their career experience, there were further opportunities to meet a wide range of female professors. The Forum's aim is to connect and network the female professors, to support the career advancement of assistant professors, to develop influence within the university and to provide role models for the younger career researchers at the university.[1] My early involvement meant that, when I began to develop the book project in earnest, I had the opportunity to contact several of the professors who I had already met during this time period.

The second influential experience was my intense work with young career researchers in the NCCR MUST initiative, through organizing scientific meetings

[1] Keller, U., & Garry, A. (March 2014). *Establishing a Women Professors Forum.* Reflections in Diversity Column. Optics & Photonics News.

such as the annual meeting and supporting other specialist schools where they had the opportunity to present their research findings. We also inform these early career researchers about potential next steps in the career and guide them toward suitable financial awards at the national level, which includes funds for assistant professorships. It was possible to see that out of those interested in an academic career, there were some already combining academic work with having a family, and also seeking suitable ways, together with their partners, for both to remain in the university research environment. This aspect of the MUST work showed the benefit to all young career researchers of seeing the different ways possible to become professor, combining all aspects of life, particularly given that it is a career path with challenging demands.

The third influence on this book came from the experience of establishing a network and support structure for younger MUST women scientists, mainly PhDs and postdocs. We include also alumni members who have moved on, either within academia or working outside academia in a range of positions in industry, consultancy, and public services; we have a membership of over 100 women in 2019. Given that women are a minority in the MUST disciplines of physics and chemistry, we wanted to encourage them to remain in science: by designing specific career workshops, offering access to female role models and by advising them individually. My work with the young women across our network gave me deeper insight into the choices and challenges of young female academics taking next steps in their careers: managing career transitions, applying for grants, developing a research reputation through publications, dealing with dual careers, and for some, becoming a parent whilst continuing in a career.

Through other MUST activities, I also spoke with younger women at undergraduate and master's levels, and was told how much they value the opportunity to gain insight into how female professors (and also women moving into other career sectors) have managed their careers. Indeed, many young women are organizing their own university societies and events to increase the possibility of hearing from experienced women, and enable them to expand their vision of future career options. This is particularly important when your chosen career lies in science, technology, engineering, and mathematical subjects in academia or industry, where women are in the minority the higher one goes in the career ladder.[2]

Thus the two goals that are central to the book emerged: to make visible and celebrate the work of female professors at all career stages; and, secondly, to do this by finding a way that would also benefit the generations to follow. These aims influenced the methodology that was developed to present the life experiences of the women professors in the book.

[2] There are 10 self-established associations at ETH Zürich for younger women students and researchers across disciplines such as engineering, computer science, mechanical and electrical engineer, physics, mathematics, and social sciences. A full list can be found on the Equal Opportunities Office Website: https://www.ethz.ch/services/en/employment-and-work/working-environment/equal-opportunities.html.

How the methodology evolved

As is often the case, the underlying concept for the book grew and developed with time. When I began the project, I could not be sure that busy women professors would be interested in participating in a project where they were asked to speak about the paths that led to their current positions and the factors which gave them the aspirations for this career. There is pressure on the few female professors in academia, not only to pursue a successful career within university environments but also to participate in activities and outreach as female role models who enthuse and inspire the next generation, particularly young women. This expectation can mean extra responsibilities particularly if you work in a discipline where there are very few women. Donna Strickland, a 2018 winner of the Nobel Prize for Physics, and only the third women recipient in the 117-year history of the awards, was surprised to be confronted with the intense focus on her rarity as a successful female physicist, rather than on her significant scientific achievements in winning the Nobel Prize.[3] However, I discovered, to my great pleasure, that one professor after the other was prepared to talk with me about their working lives. This book would not have been possible without their time and generosity.

The next step was to develop an interview process or technique. My early instinct was to create a method that enabled each professor the freedom to speak about her experience naturally, which would also produce a sense of how their careers evolved over time. I wanted to understand the origins of the professor's interest in their topic, who inspired them to study, and how, over the years, their career path came into being. I designed a series of general questions (see Appendix 1), which would uncover how it all started and what or who influenced the steps taken. These questions were then adapted to each professor's circumstances, using information I was able to find out about them in advance, from a variety of sources. In reality our meetings evolved organically and the questions were used as a prompt to a natural discussion, rather than a framework to be followed tightly. They were a potential checklist to keep the unfolding stories on track, and also to garner the professor's positive experiences, potential challenges, and general advice for anyone considering this career. This methodology was used with all the interviews conducted for this book, and each meeting took between 1 and 2 hours.[4] In some cases, there were subsequent meetings, especially if there were significant developments in a person's life or career.

This became the nucleus of my vision: to organize in-depth conversations with female professors from a diversity of origins and life experiences, and to compose then a book featuring their careers. The next steps went relatively smoothly: the first contact with Elsevier happened at the EU Gender Summit in 2014, some

[3] *I see myself as a scientist, not a woman in science*, Donna Strickland interview, The Guardian, 20 October, 2018. https://www.theguardian.com/science/2018/oct/20/nobel-laureate-donna-strickland-i-see-myself-as-a-scientist-not-a-woman-in-science and Prof. Athene Donald (Master of Churchill College and Physics Professor, University of Cambridge reflects on this point in her blog, *To be or not to be a Role Model*, November 4, 2018. http://occamstypewriter.org/athenedonald/.

[4] Two interviews were conducted via Skype with professors who have moved abroad.

correspondence with the publisher followed, then I submitted my book proposal, which was accepted.

In the earliest stages of the project I had thought only to gather the information together and then extract and share commonalities between all the female professors' careers; but over the time spent talking with these women, and hearing about their diverse choices, their research questions, their enthusiasm for teaching and students, and for working in an academic environment, I realized that the professors were the central part of this narrative. I decided to make their personal voices the communication channel for their experiences—I would write each meeting, or meetings, as a live conversation. The reader could experience our discussion in a format where the professors told their stories directly to me. That is the first factor, which influenced the shape of the conversations with the professors: to let their own voice be heard in an authentic way.

The second factor which influenced this decision was the fact that, currently, the stories of career women are often dominated by describing their working lives through their minority status, or low numbers in the higher levels of academia, and wider careers in general. This topic can take over the way women's working lives are represented. They are often analyzed and categorized by descriptions of how unusual and groundbreaking they are, how different to the norm, without giving enough space for how they got there and what they want to achieve in their fields and why.

The idea of creating "Inspiring Conversations" with the key purpose of hearing directly from these expert and distinguished women was born. This method provides a human voice behind the career, and also a direct sense of what it mean for each individual woman to become a professor.[5] Once I decided on this method, I still had to discover whether it would be acceptable for the professors involved, which meant that producing the conversations that appear in this book involved a further two steps.

In the next step I wrote a draft conversation for each person, in such a way as to encompass the free flow discussions that we had in our meeting(s); in doing so I created the first step of the professor's voice. Then the creation of the conversations became interactive. I wanted to respect what they chose to include in the conversation, and what should be removed. The professors were given the opportunity to check for accuracy, add important details, and remove aspects that would remain private, and make the final decisions about what would be published. This was a final step in respecting their voices and how they wanted to portray their careers. It is clear that these conversations are snapshots of particular periods in the professor's life and that their perspectives will evolve with further experiences, but they contribute significant value through their openness about moments of uncertainty, other paths taken before finding the way, the important and generous people who influenced them, and the supportive partners and families who grew with them along the way.

[5] Mary Beards' book, *Women and Power*, published in 2017, includes an essay, "The Public Voice of Women," about the historical treatment of women's voices confirmed my instinct on the importance of hearing directly from the professors.

The professors who contributed

Two factors were important in my decisions to approach the professors who would take part in the book. First, I wanted to talk to women who had already been appointed as a professor; because it mattered to learn from them how they had made this significant step. This is particularly pertinent, given that the percentage of appointed female professors in academic institutions is still low, standing at 21% across all disciplines in European Universities.[6] A second factor in the choice of professors, was the importance of including women professors from each step of the career: assistant professors (with or without tenure-track), associated professors with tenure, and full professors. By including a range of career stages, a wider sense of the experiences and challenges at different levels of the professorial career is given, and a deeper insight into how these career transitions were made. On the European continent being appointed as a professor means that, even at the early stages, you will have an established research reputation and the position will include funds to set up a research group, a laboratory (if you are an experimentalist), and you will take on teaching responsibilities wherever you are appointed. For full details of the different types of professorship appointed in Switzerland, see the Glossary.

Throughout the whole period of the book project, I met with many female professors in a range of different circumstances: at conferences, scientific meetings, and career workshops, to name a few. These discussions added to the perspective of the career landscape. For the book, I spoke formally with 27 women professors. As is normal in any long process, not everyone agreed to be included in the final version: the timing was not right, their perspective had shifted and they did not want to revisit the discussion, or they had moved countries and positions. The book has 23 conversations with professors aged from mid-30s to over 70 years old. The women come from eleven different countries, from different continents and have a wider range of nationalities. During the last four years, some of the participants went through career progression from assistant professor (with or without tenure-track), to associate professor, and even to full professor. Six became tenured over the time period of the book, and all are now tenure-track. The conversations, as planned, give a strong representation of transitions made in university environments. Nine of the professors have long careers, which means that the reader gains insight into the expansion of roles as they progressed to the leadership level in academia. The majority of the professors have positions at ETH Zürich, though one is long retired. There are representatives from the University of Zürich, EPFL, Lausanne, and the University of Bern, while two professors moved abroad to pursue their careers, and still agreed to contribute to the book, adding insight into how mobility can be a factor in finding the next steps of working life.

[6] The *She Figures* 2015, European Commission, Brussels, 2016. See also Figure 1, Section 2.

The majority of contributors to this book work in a range of scientific and engineering disciplines, but there are also contributors from related fields in psychology and social science. It gives me great pleasure, in the next chapter, to present what they told me about their aspirations, lives and careers.

Twenty-three conversations with

1. **Natalie Banerji** – Department of Chemistry and Biochemistry, University of Bern, Bern, Switzerland
2. **Eleni Chatzi** – Department of Civil, Environmental and Geomatic Engineering, ETH Zürich, Zurich, Switzerland
3. **Emanuela Del Gado** – Department of Physics, Georgetown University, Washington, DC, United States
4. **Rachel Grange** – Department of Physics, ETH Zürich, Zurich, Switzerland
5. **Stefanie Hellweg** – Department of Civil, Environmental and Geomatic Engineering, ETH Zürich, Zurich, Switzerland
6. **Ursula Keller** – Department of Physics, ETH Zürich, Zurich, Switzerland
7. **Salomé LeibundGut-Landmann** – VetSuisse Faculty, University of Zürich, Zurich, Switzerland
8. **Ulrike Lohmann** – Department of Environmental Systems Science, ETH Zürich, Zurich, Switzerland
9. **Marloes Maathuis** – Department of Mathematics, ETH Zürich, Zurich, Switzerland
10. **Isabelle Mansuy** – Department of Health, Sciences and Technology, ETH Zürich, Switzerland; Department of Medicine, University of Zürich, Zurich, Switzerland
11. **Paola Picotti** – Department of Biology, ETH Zürich, Zurich, Switzerland
12. **Ursula Röthlisberger** – Institute of Chemical Sciences and Engineering, EPFL, Lausanne, Switzerland
13. **Clara Saraceno** – Department of Electronics and Information Technology, Ruhr Bochum University, Bochum, Germany
14. **Maria Schönbächler** – Department of Earth Sciences, ETH Zürich, Zurich, Switzerland
15. **Olga Sorkine-Hornung** – Department of Computer Science, ETH Zürich, Zurich, Switzerland
16. **Nicola Spaldin** – Department of Materials, ETH Zürich, Zurich, Switzerland
17. **Elsbeth Stern** – Department of Humanities, Social and Political Sciences, ETH Zürich, Zurich, Switzerland
18. **Shana Sturla** – Department of Health, Sciences and Technology, ETH Zürich, Switzerland
19. **Brigitte von Rechenberg** – VetSuisse Faculty, University of Zürich, Zurich, Switzerland
20. **Katharina von Salis (Retired professor, since 1990s)** – Department of Earth Sciences, ETH Zürich, Zurich, Switzerland
21. **Effy Vayena** – Department of Health, Sciences and Technology, ETH Zürich, Switzerland
22. **Sabine Werner** – Department of Biology, ETH Zürich, Zurich, Switzerland
23. **Marcy Zenobi-Wong** – Department of Health, Sciences and Technology, ETH Zürich, Switzerland

Inspiring Conversations with Women Professors. DOI: https://doi.org/10.1016/B978-0-12-812346-1.00001-9
© 2019 Elsevier Inc. All rights reserved.

Prof. Natalie Banerji (Austrian)

https://banerji.dcb.unibe.ch/
FemtoMat Lab, Department of Chemistry and Biochemistry, University of Bern

Natalie Banerji

Biography

Natalie Banerji is currently a full professor of Chemistry at the University of Bern and leads the FemtoMat group. Her research interests include the study of organic and hybrid materials using ultrafast spectroscopic techniques, in view of solar cell and bioelectronic applications. She studied chemistry at the University of Geneva and obtained her PhD in physical chemistry in 2009, under the supervision of Prof. Eric Vauthey. She then moved to the University of California, Santa Barbara (United States), to work on organic solar cells during a postdoctoral stay with Nobel Laureate Prof. Alan J. Heeger (2009–11). In 2011 she started her independent research career in Switzerland at the Ecole Polytechnique Fédérale de Lausanne (EPFL) with an Ambizione Fellowship by the Swiss National Science Foundation (SNSF). She moved to the University of Fribourg in 2014 as an SNSF stipend professor, was subsequently nominated as tenured associate professor in 2015, and presided over the Chemistry Department in Fribourg from 2016 to 2017. She moved to Bern in 2017.

Research area

In the FemtoMat group, we investigate what happens on the ultrashort time scale and ultrasmall length scale in organic and hybrid materials in order to induce macroscopic function in electronic devices, and how this can be optimized. We use a complementary palette of techniques combining time-resolved spectroscopy, pulsed photocurrent methods, terahertz experiments, Stark-effect spectroscopy, and device testing.

Honors and awards

2016	ERC Starting Grant
2015	Grammaticakis-Neumann Prize by the Swiss Chemical Society
2014–15	SNSF Professorship Stipend
2011–14	SNSF Ambizione Award
2009–11	Early Postdoc Mobility Award

Conversation with Natalie Banerji, October 24, 2018

Were you interested in studying chemistry from an early age?

Actually for a long time I wanted to be a Vet and my school choices were based on that. I loved science, and I had even chemical experimental sets at home, but my choices were made with the purpose of applying for veterinary medicine. I went to an international school in Geneva, which meant studying for the International Baccalaureate. To apply for Vet school, you needed to study all three natural sciences—biology, physics, and chemistry—and I chose that option with a concentration on higher level biology. However, at 16 years of age, I spent a week working in a Vet's practice and discovered that seeing what the job actually entailed, I never wanted to do it again. It was not what I'd imagined and I did not enjoy it at all.

How did you cope with this sudden change of direction?

It's something I've learned about myself over the years, that if a situation is not right, I always find a way to find an alternative as soon as I can. This time, because I was specializing in biology at high school, I thought that biology might be the best option for a degree course, but I sought advice first. I talked with a professor of biochemistry at the University of Geneva and explained that I was interested in molecular biology and he recommended that biochemistry was a better choice to fit my interests. This discussion influenced the next step.

You studied chemistry and biochemistry at the University of Geneva. How was that experience?

I took a degree in chemistry and biochemistry and found that during the first 2 years, while studying the fundamentals, I preferred chemistry and focused on that for the rest of the 4-year diploma. I did well on the courses, but I wanted to move into industry rather than studying for a PhD. When it became clear that to work in industry it was better to have a PhD, I changed my mind. I was offered PhD positions in a number of groups in Geneva, because I had good grades in my diploma. I also met my partner at this time, which confirmed my decision to stay in Geneva, rather than moving elsewhere.

Your next steps were now clear. How did you experience your PhD period?

In reality nothing was clear for some time. I joined an organic chemistry group as a PhD student, but did not enjoy the research in this area and left after a few months. This made it more difficult to find another group as a PhD student. It was only after speaking to Prof. Eric Vauthey, and to members of his group, about his research in physical chemistry and ultrafast spectroscopy, and finding it very interesting, that a new option opened. I agreed to start on a trial period, to see if it would work out, and then was officially hired after a few months, to work on photo-induced electron transfer reactions. I really liked the topic, but was still unconvinced that I wanted an academic career. So, in my third year, I worked 50% on my PhD and 50% as a teacher in the international school where I obtained my Baccalaureate. I wanted to check whether high school teaching was a career I should consider. I was thrown in at the deep end, found dealing with teenagers quite stressful and unfulfilling, and overall I did not enjoy the experience. I committed seriously to completing the PhD and went back to full-time research.

You went next to the United States as a postdoc, had you committed to academia at this point?

I was still not convinced, but I thought the experience abroad in the United States was an excellent opportunity for broadening life experience too. I applied successfully for an SNSF postdoc mobility award, 6 months before my PhD defense. My plan was to return to Switzerland after a year and study for an MBA. However, the time in the United States changed my opinion about a research career and I began to take it seriously as an option.

What was it about the United States experience that crystalized your choices?

I joined the group of Prof. Alan J. Heeger (Chemistry Nobel Laureate 2000) at the University of California, Santa Barbara. It was a really great experience, the group was very good, and I was able to work independently and find my projects. I started to identify my own research direction, which was exciting. I had a great time research wise and socially; we worked hard, socialized together and had a lot of academic freedom. This is what attracted me to a research career, the potential for independence and making an impact on a field. A factor that shaped the next step though was that I wanted to return to Switzerland, where I saw more funding opportunities for young researchers and where my boyfriend had a permanent job. We had already lived separately for almost 2 years.

How did you make the move back to Switzerland?

I approached Prof. Jacques-E Moser, from EPFL, Lausanne who I was introduced to when he sat on my PhD commission. He offered me a regular postdoc position in his group. While still in the United States, I had already applied for the SNSF Ambizione Award, which gives young researchers funding to establish their own group and I received this award a few months after my return, which was extremely a positive news.

You established your group in the lab space of Jacques Moser?

Jacques Moser was very supportive and he gave me the space to develop my research independently, which gave me the opportunity to discover and develop my research direction. It was a challenging time, initially, because I came from the

United States where I'd been embedded in a large group and socially integrated there, and now I was in a new position where I was more isolated, because I had taken the next step in academia. It was harder to make new friends as I established my group, plus my home base was near Geneva and not in Lausanne. I concentrated on my research work, produced lots of results and publications, and started to receive many international invitations to conferences. I achieved some very well received papers, based on thorough science, and built an international network of collaborators, which were essential steps in grounding my career.

Did you now have plans for your future?

Because it was clear that I wanted to be a professor, in my first year at EPFL, I started to apply for professorships all over the world. I would have 1–2 interviews per year, which was great practice and I often came second on the list, which was positively reinforcing. It was affirming to get close to a job offer, but at this point I was also glad that I did not have to consider a big life change such as moving to another country.

In the third year of my Ambizione project, a number of professorships came available in Switzerland. I applied for all the positions, and also for an SNSF stipend professorship, which could be based in a host university. I came second on the list for a University of Zürich position, and then I was offered an Associate professor position at the University of Fribourg, and at the same time, I was awarded the SNSF professorship, also attached to Fribourg. I had the SNSF professorship for 1 year, and was awarded tenure in 2015. The SNSF award boosted my funds substantially enabling me to set up and build my laser lab in Fribourg, which is a very expensive undertaking.

This means that within 4 years of your return from the United States you had achieved everything you wanted. How would you describe your experience of this?

The challenge at this point was that I was inexperienced, and I had joined a small department where, full of enthusiasm to contribute, I said "yes" to everything. I was on every possible commission and, after a year, I was made head of department. I also successfully applied for an ERC starting grant at this time. There was so much work to do that it became very stressful and my science was suffering because of the university commitments. I was drowning in work, and I actually did not realize how bad it was at the time. I learned a lot from this experience about what I need to develop for the future.

You are now a full professor at the University of Bern. How did this come about?

I saw the job advert for the professorship in Bern, but decided not to apply because I was new in my position in Fribourg. However, I was approached by members of the University of Bern who encouraged me to put in an application anyway. I thought it was a long shot, but I applied. I came second on the list, and put it out of my mind. Then, just before Christmas in 2016, I learned that the negotiations for the first candidate had fallen through, and I was offered a full professorship. I decided to take the position.

I knew that moving my lab would slow my research progress initially, but that on the medium term I could build a strong position based on excellent lab space and research means in Bern. I had always expected that I would not stay in Fribourg, and that there would be a next position. This move has taken me to my forever position earlier than I expected. The last year has been full of challenges in order to get to this point. Many changes happened in parallel, including in my personal life. I feel confident and relaxed now that the lab has moved to Bern and actually was operational faster than I foresaw. My group has got much bigger and is still growing, I have four postdocs, six PhD students and open positions. My ERC grant application was also successful, providing a boost to the group.

How would you summarize the positive experiences in your career to this point?

What is really great is that I am passionate about my research, which means that I am always working on something I like to do. I am also part of the organic electronics community, which is a great community with a lot of international contacts. I really love working with students and with my research group, we have a great group of people working together. Being a professor means that you have the space to develop independently and you are your own boss, which is also very positive.

For me to become a full professor at 36 years has been wonderful. I feel that my life has become more settled, I feel respected by colleagues and I don't feel the pressure to say "yes" to everything. As a full professor you can think longer term with the research, reflect on the results, examine them and take more time to build a research program.

What would you identify as the challenges?

One challenge of being a full professor is that I feel there is a lot on my shoulders, because I am responsible for many people, scientific projects, teaching, and other tasks. I think it is possible to feel constantly guilty and under pressure, because you never get through everything that accumulates on your desk. You have to deal with things that you were never trained for, including a lot of management tasks. In addition, there can be a lot of intense traveling, which means you have to delegate and you have to fight for the research time. Specifically, in my case, I feel it challenging to be a physical chemist, which means that I neither fully "belong" to the chemists nor to the physicists. I would say it is more challenging to be a physical chemist in my department than to be the only woman professor there. In the earlier stage of my academic career, while taking the steps to a full professorship, I found it challenging that you are given a lot of work and responsibility outside of your research activities, and it is very hard to refuse at this stage. You need to find a constructive balance at this point.

What were the factors that supported your career?

Of central importance was that I worked very hard to get all my results. It is just as important to have an independent personality, or develop this, where you learn to trust your own gut and go for it. Being your own boss means that no one really tells you what you need to do. I also, definitely, had a lot of good luck, being at the right place at the right time. For example, the position opened up in Bern at a very good time and, in the end, I was offered the job. Another important factor is that I have learned to find perspective and a thick skin through more difficult times. In a research career, constantly being evaluated (by reviewers, job panels, grant commissions, students, etc.) and often being criticized or rejected, is part of the job. I learned that sometimes a situation can feel impossibly difficult, but things always work out in the end no matter how bad they feel at the time.

What advice would you give to younger people interested in an academic career?

It's important to be proactive and look early for positions, even if no one is encouraging to do so. Secondly, you need to really reflect on what you want to do with your life. Examine the job of a professor realistically and see if it is the type of job you would want, balancing the gains against the challenges. I have observed that many people are put off the career because it needs a lot of effort, nights and weekends. I am totally sure that this job fits me as a career and I am happy about it, even when there are very busy times and it can be frustrating. I wouldn't do anything else!

Any final thoughts on your influences?

I had some important role models and collaborators in my wider field and scientific community, among them women academics. They helped me to know how to behave, or fit into, my community. They also talked very positively about my research, leading to many invitations to conferences and me developing an extensive international network.

Overall, I would say that it is not a piece of cake to achieve an academic career but, at least for me, definitely worth all the effort!

Prof. Eleni Chatzi (Greek)

http://www.chatzi.ibk.ethz.ch
Structural Mechanics, Institute of Structural Engineering, Department of Civil, Environmental and Geometric Engineering

Eleni Chatzi, "ETH Zurich/Giulia Marthaler"

Biography

Eleni Chatzi is currently an associate professor and the Chair of Structural Mechanics, Institute of Structural Engineering, Department of Civil, Environmental and Geomatic Engineering (DBAUG), ETH Zürich. She obtained her diploma (2004) and MSc (2006) in civil engineering, with honors, from the Department of Civil Engineering at the National Technical University of Athens (NTUA). In June 2010 she obtained her PhD degree with distinction from the Department of Civil Engineering & Engineering Mechanics at Columbia University, New York. In 2010 she was hired as the youngest assistant professor in ETH, and was promoted to an associate professor in 2017.

Research area

Her research couples novel simulation tools with state-of-the-art monitoring methodologies for smart infrastructure assessment, with the goal of providing actionable tools able to guide operators and engineers in the management of engineered systems. A key aspect of her research lies in extraction of quantifiable metrics that are indicative of structural performance across the component, system, and network levels. Her research interests lie in the area of structural health monitoring, with a strong focus on problems lying beyond the commonly adopted assumption of linear time invariant systems. Her research spans a broad range of topics, including applications on emerging sensor technologies and structural control, methods for curbing uncertainties in structural diagnostics and life-cycle assessment, as well as advanced schemes for nonlinear/nonstationary dynamics simulations.

Selected awards and honors

2018 ERC Proof of Concept Grant
2018 Marquis Who's Who in the World
2016 Recipient of the Francis Ogilvie Young Investigator Lecture award, awarded by the Center for Ocean Engineering, Massachusetts Institute of Technology (MIT)
2015 ERC Starting Grant
2010 Distinction from the Fu Foundation School of Engineering and Applied Science at Columbia University for the Doctoral Dissertation
2009 The Mindlin Award—Department of Civil Engineering and Engineering Mechanics in Columbia University
2006 Fulbright Scholarship

Conversation with Eleni Chatzi, October 23, 2018

Where did you begin your journey to becoming a civil engineer?

My parents encouraged and supported me to get a good education. My father is a physicist, which might have contributed to my development as a student who excelled in math and physics. Like several top students in my year, and after scoring fifth on the nationwide entrance examination, I was encouraged to apply for the prestigious National Technical University of Athens. I was always drawn by fundamental principles and was very pleased to find these aspects strongly present in the civil engineering curriculum. It is these fundamental principles that drive my research to date, as I try to move on the interfaces between civil and further engineering disciplines. My work relies on the use of sensors for better understanding of built structures and engineered systems, but is at its core driven by the theory describing the underlying processes. As I develop my research on use of methods and tools from the civil, mechanical, and electrical engineering disciplines, I am by now convinced that innovation lies on the interface of disciplines.

I graduated with a diploma degree in civil engineering from NTUA. The first 2 years focused on fundamental science, while years 3–5 introduced the more applied "engineering aspects" of the discipline. The percentage of women was at that time low at 15%–20%, but I understand this rate has increased in recent years, tending to a more balanced distribution, and reflecting the potential and interest of women for engineering. I followed my diploma with a 2-year master's course in structural analysis and seismic design.

What drove you to move in an academic direction rather than to industry?

Perhaps a main driver in my academic development came from the influence of Prof. Valsis Koumousis, who was my diploma and master's thesis supervisor. He gave me a feel for research and instilled in me a curiosity about science, which I wanted to pursue further. He was an inspiring figure for me, with whom I maintain a close contact. Some people just make a mark on our lives. He encouraged me to apply for a PhD in the United States, and I applied successfully for a Fulbright

Scholarship. This was to be a huge move for me, changing countries and continents. I applied to a number of universities, but eventually chose Columbia University, after discussing with Prof. Andrew Smyth, who was a young professor at Columbia. He reached out to me, and gave me an idea of the research topics I would work on, introducing me for the first time in the field of structural health monitoring. It was an exciting new field and both the concept behind the field as well as the evident supportive structure within the group convinced me to study there.

Had you already envisioned that you would become an academic?

No, I was never someone who made becoming a professor a goal. I grew my interests as opportunities came along. It was really exciting, and a life-changing experience, to make the move from Athens to New York. Studying in New York is a fascinating experience on all levels and New York is a city that caters to every taste. This made my transatlantic transition a smooth one. I studied on the main campus of Columbia University, ironically, in the ugliest building on campus, as is often the case for civil engineering department buildings.

I had a wonderful PhD supervisor. He allowed me to work independently, but also gave me critical feedback and suggested new directions of research without being invasive or over-controlling. This created a great feel for the ambience of research communities and a very positive vision of the academic environment. I slowly discovered the academic world through the best of circumstances and I've used this experience in my own supervision methods, relying on the good examples I have to fall back on. As a result, I took steps in my career quickly, but this was not preconceived; it rather came as the result of having a good support system, which empowered me to take the next step.

You became an assistant professor (AP) straight after your PhD. What made you take, and achieve, this step immediately?

It was my PhD supervisor who suggested that I should apply directly for assistant professor positions, rather than postdocs, as I had already built a strong research profile. I applied to 5–6 institutions and received three job offers, one of which included ETH Zürich. I was dumbfounded by the offer coming from ETH. It had not occurred to me to return to Europe, and most of my applications were in the United States, but I came upon the advert for ETH and applied. It was clear at the ETH interview that it was a unique opportunity with great support, excellent resources and an ideal environment for research. I was delighted to accept the offer.

What happened once you joined ETH Zürich in 2010 as an assistant professor?

I was hired as a non tenure-track assistant professor (supported by the Albert Lück Stiftung), with a 4 plus 2-year contract. Everything had happened so quickly that I took the position without fully realizing what non tenure-track means; in the United States almost all positions are tenure-track. Once I came to ETH, I understood that the position could imply a dead-end. A permanent position in my department is typically tied to the retirement of a professor, or chair, with the reappointment usually carried out in line with the research of the previous chair. Naturally, this need not necessarily fit with the scope of an existing non tenure-track AP position. However, I learned to see my professorship as a stepping stone, and a great means for supporting forefront and independent research, and, in the third year, I started applying for tenure-track positions in other institutions.

At this point, I received some very good offers from universities in Europe and the United States, where much of the good work in my field is pursued, and where the academic environments further impressed me. I had decided to accept one of these offers, but when announcing my decision to leave to the ETH President at the time, he decided to make a counter offer for staying at ETH. He discussed this option with my department, and I proceeded to personally talk this potential offer through with my colleagues in the department in a short space of time. It was for me essential to ensure that my stay in the department, as a tenure-track appointment, would be positively perceived. This procedure, which naturally came with multiple uncertainties, was an intensive time for me. My colleagues gave the go-ahead to the counter offer, which was hugely positive experience. In 2014 my position transferred to a tenure-track assistant professorship and I achieved tenure in 2017. In this process, I would like to acknowledge the Albert Lück Foundation, which shares my vision for research, and continues to generously support my position to date.

Everything happened very swiftly in your career. Did this have an impact on you?

My career has been very intense, I was hired as an assistant professor at 28 years old—the youngest AP hire in ETH at the time—and I was 32 when I found myself looking for a next position, and shortly after shifted to a tenure-track professorship. My ability to cope with such an intense process is largely owed to the support I received from personal and professional communities, as well as the guidance of colleagues I regard as mentors. Their advice was essential to keep me grounded at a time of strong psychological pressure.

What did you learn from this experience?

I learned that it is important to deal with challenging situations in an organized, yet timely manner. It is not prudent to act impulsively, but at the same time it is also not wise to remain in a transitional phase for too long as this leads to delays in research output and has an impact on the relationship of a supervisor with their research group. In a way, the non tenure-track position is one of these transitional phases. I was given a lot of sober advice on this topic by my departmental mentor at ETH, Thomas Vogel, as well as further colleagues from both within and outside ETH. I listened carefully to their inputs and approached my non tenure-track experience with appreciation, but also by planning for what comes next. Perhaps one of the toughest times in my academic career so far was the decision to remain at ETH. The shift to a tenure-track appointment, allowed me to regain a feeling of certainty, and a level of control in my career, as I felt that the next stages would depend on my own ability to deliver and my documented research output; two parameters which really depended on my performance and not circumstances. In 2015 I had a tremendous boost when I was awarded the ERC starting grant for research in the smart monitoring and assessment of wind turbine infrastructure—it was a real incentive to embark on exciting research with new ideas and novel methods and tools.

During the tenure process there is a feeling of responsibility toward the students and researchers that one employs. Careful planning needs to be put in place so as to not endanger or negatively impact the careers of the group members. Since 2017, having achieved tenure, I can confidently plan for next steps and build up my group. Thanks to the generous support of ETH, and the start-up package that comes with a tenured appointment, I am now building up a new and unique experimental facility. This will be smart lab for dynamically testing structures under changing environments—a piece of reality in the lab. I now lead a highly interdisciplinary group of 18 students and researchers of diverse ethnicities and backgrounds, which is also about to grow again due to recent EU and Swiss grants.

What would you say is central to your experience of being a professor?

I would say that my research activity characterizes me as an academic. I have by now formed specific expertise, which places me among experts in the domain of structural health monitoring. I use sensors to monitor the health, or condition, of structures and engineered systems, with applications spanning bridges, buildings, aerospace structures, wind turbines and even railway vehicles and tracks. A wonderful aspect of this research is that we work at the interface of engineering disciplines such as civil, mechanical, and electrical engineering, as motivated by our coupling of sensors with structures. By fusing technology with appropriate data processing tools, my research balances the theoretical with the applied. In fact, you will often find us on-site installing sensors onto structures ourselves, sometimes in zero temperatures! On the other hand, we also rely on the use of theory and computation for simulation of the systems we monitor. In this respect I have the pleasure to coordinate the PhD Program on Computational Science, a joint venture between ETH Zurich and the University of Zurich, which brings together students from seemingly disparate disciplines that are however interlinked via the use of similar computing methods and tools. Such efforts spark academic cross collaborations across disciplines. I believe that this type of cross-disciplinary research may change the shape of universities in the future.

What factors had a positive impact on your career?

I have already hinted at this, but I do believe that a significant factor for me was the constructive and inspiring support I received from my supervisors and mentors throughout the early stages of my career. A second factor lies in being introduced early into this highly active research domain, establishing myself as a core member of the monitoring research community.

Did you have any identifiable challenges?

The somehow fast track in my career evolution, which followed a somewhat unorthodox path, was certainly a challenge for me. I was forced to mature at a rather early stage. Currently, what I perceive as my major challenge is to be a successful supervisor for my group, although I am happy to have this responsibility. Managing a highly diverse group from masters, to PhD students, to postdocs and senior scientists, forms an everyday challenge. It is a complicated dynamic, but certainly a functional one when you treat people with respect. I also like to encourage young women to join the group, because we cannot afford to lose them. I hope to open up their eyes to possibilities in engineering and academia, particularly since I realize how much the encouragement from my early supervisors helped me.

What advice would you give to young women aspiring to academia?

You should focus on what is truly important in this respect—research and teaching. Do not be discouraged by different social behaviors and ill-natured comments that could come along the way, unfortunately within the community. Such problematic behaviors form common symptoms across any kind of societal structure. I also found that I received comments on one of my early grants, with someone saying it was awarded to me due to my gender and related acceptance quota. As a woman in civil engineering, which still is a male-dominated field, I've had the chance to observe behaviors of researchers at different levels and of different genders. I tend to discern a tendency of young female researchers to perpetually defend their work, instead of acknowledging their achievements and building onto previous accomplishments. This is my advice to young female colleagues: remain focused on your vision, trust your skills, do not be discouraged by negativity, and keep in mind that it is research contribution, scholarship, and professional conduct that define an academic, thereby making us all equal.

I also find it vital to encourage female engineers to consider, and apply for, faculty positions in the engineering domain, which unfortunately is an all too infrequent occurrence.

Prof. Emanuela Del Gado (Italian)

https://physics.georgetown.edu/users/emanuela-delgado
Department of Physics, Georgetown University, Washington, DC, United States

Emanuela Del Gado

Biography

Emanuela Del Gado is provost's distinguished associate professor at Georgetown University, Washington, DC, United States. She received her undergraduate degree (Laurea in Physics, cum laude) at the University of Naples "Federico II" in Italy, where she also obtained a PhD in Physics in 2001. She was a Marie Curie Fellow at the University of Montpellier II in France and a postdoctoral researcher at ETH Zurich in Switzerland, and held visiting positions at ESPCI (France) and MIT. Before joining Georgetown University as associate professor with tenure in 2014, Emanuela was a Swiss National Science Foundation (SNSF) assistant professor in the Department of Civil, Environmental and Geomatic Engineering at ETH Zurich.

Research

Emanuela Del Gado is a theoretical physicist working on engineering motivated problems. She uses statistical mechanics and computational physics to investigate materials with structural and dynamical complexity, from model amorphous solids, gels and glasses, to new green formulations of cement; nanoscale structure and mechanics of cement gels; self-assembly of nanoparticles and fibrils at liquid interfaces; biomimetic coatings and mechanics of tissues; mechanics and rheology of gel networks.

Awards and honors

2018	Royal Society of Chemistry Fellow
2018	Chair Paris Sciences, ESPCI Paris
2017	MIT—CEE C.C. MEI Distinguished Speaker
2017	Provost's Distinguished Associate Professor, Georgetown University

2016 Journal of Rheology Publication Award
2010–15 SNSF Professor
2002–04 Marie Curie Fellow

Conversation with Emanuela Del Gado, June 30, 2015

The excitement of finding solutions to challenging problems, and developing something new, which will contribute to many fields—that's what drives me in science.

When did your interest in physics begin?

I always wanted to understand the world, but I did not originally study math and physics. I went to an academic high school in Italy, where I studied Greek, Latin and Ancient Literature. In the Italian *Liceo Classico*, ancient Greek and Latin are taught as dead languages and you work at translating ancient authors. I loved it, it was a lot about logic, very mathematical in the end, in the sense that we had to practice finding solutions to complex problems. I also studied Philosophy and, when I studied 20th century philosophy and its connection to science, I became very interested in Physics. I was always attracted to challenging subjects and was driven to study them. I took my undergraduate degree in Physics at the University of Naples Federico II.

What made you continue to a doctorate?

In reality when I started my PhD I did not know whether I really wanted to continue doing academic research. I was worried that the part of me that needed creativity in my life would not be satisfied. I totally fell in love with research, when I discovered that a lot of it is about creativity. The problem initially conceived by someone else transforms into your problem, and you can develop and test your own ideas. When you see the connections between the initial problem, the subsequent solutions and the new emerging questions, there is no job that is better than this. Nothing usually works at first, you have to dive into the problem, climb the mountain and then get over the top to find the answers.

As an undergraduate, I always tried to take the most difficult courses. It gave me a strong grounding for later and made subsequent research projects easier. I had inspirational teachers especially in theory and this probably pushed me in that direction. For my PhD I worked in theoretical physics, but I focused more on computer simulations. Now at Georgetown University, in the United States, my research uses a lot of simulations and we work in close contact with experimentalists. While I like to think in abstract ways and to address fundamental questions, I want to make sure that I understand the physical phenomena we are trying to unravel. For this, I need to work close to experimentalists and sometimes I think I would have liked to be an experimentalist myself. I really need this combination between theory and experiments, where theoretical physics is connected to real problems and ongoing experiments.

What were the next steps in your career?

It was not straightforward for me. I wanted very much to explore other situations and to test my capabilities in different contexts. I applied for postdoc fellowships abroad and was successful with my application for a Marie Curie fellowship, which meant that I could move as a postdoc to France with independent funding. I had conceived my project together with my supervisor, so I was very much into it. I believe it's really important, to develop as a scientist, to have the experience of changing country, working with different people, and having different supervisors.

Your next step was to go to ETH Zürich as a postdoc?

After the Marie Curie Fellowship in France, I went back to Italy for a short period, but I felt very strongly that there was more that I wanted to learn and explore. So, I decided to leave and applied for a number of different positions again abroad. Then an opportunity came available at ETH Zürich as a postdoc in the Department of Materials. It was a really great place to work, the research environment was very stimulating and the city is really nice.

At ETH Zürich I learned how freedom and resources are important factors for successful research. When I applied for a SNSF Professorship, my research plan was quite unusual, doing material physics in a Department of Civil Engineering, because I wanted to work on cement. It was great when the funding body supported me to do something adventurous, which had never been done before.

I know that you had a two-body or dual-career situation over this time. Could you explain how you handled this?

I met my husband in France, he's also a scientist and at that time he was working in United Kingdom. His career was already established there and so I thought about moving to the United Kingdom as well, but the SNSF position at ETH, which meant being based in Switzerland for a fixed term of 6 years, was particularly exciting for me, because of the originality of the research topic and the unique opportunity to start something new. I did not want to give up this opportunity and working at ETH brings so many possibilities. Therefore, we went on working in different countries for a few years, without finding a better solution for our family life in Europe. This has been possible at Georgetown University, in the United States.

Were you concerned about finding the right position and solving the two-body problem?

My first priority was to be doing the research I wanted to do. However, it was clear to me that the next move after ETH had to be a tenured position. Our dual-career challenge turned out to be very difficult to solve in Europe. My impression is that, from an academic point of view, it is still very tricky to discuss dual-career problems, especially at an early career stage. Being two physicists working in related areas, the United States was a better option, also because of the bigger job market and because they have more proactive solutions to the dual-career problem.

I had heard of a new initiative at Georgetown in the area of "soft materials," and the university advertised a series of open professorship positions. We decided to explore the possibilities there, and we were invited to visit the university. Georgetown University is based in Washington and we found the location, the scientific and intellectual environment very interesting and stimulating. In 2014 we both joined the new established Institute for Soft Synthesis and Metrology (incorporating material design, materials, physics and chemistry).

I was hired as associate professor with tenure and I finally combine my academic career with a fulfilling personal life. I am very happy about the research collaborations and the working environment at Georgetown University. It is a great place, where people worry about gender balance and there is a culture of collaboration and inclusivity.

Is there a difference now that you have a tenured position?

I can make long-term research plans and take more risks with research ideas. Now I can think of the work on a wider scale and better identify where I can make progress in my field.

When I finished my PhD, it was not obvious to me that I wanted to become an academic, I was worried about having supervising responsibilities and not interested in jobs that involved teaching. Then I began to see teaching as part of the job, or package, of being a professor. I enjoyed supervising and marveled at the transformation that could take place as a person realized that it was their own problem they were investigating and that they could solve it. This is exciting. I really enjoy working with younger people, students and postdocs. It is great to work at a university, if I were at a research institute I would miss this connection.

Have you experienced any challenges as a woman in your field?

As an undergraduate student, the gender ratio was 50:50. Afterwards it was a bit more complex, because it became mostly a male environment and I started to notice personal comments about makeup and clothes. I remember trying to talk with a male colleague about feeling stressed after giving a talk, and that he asked me whether the issue was that my makeup had melted!

As a young female faculty, I experienced that, if you are not assertive enough people tend to underestimate you. Then if you are assertive, they tell you that you are aggressive. I also realized that there is a lot of unconscious bias in academic life and that women are often exposed to unfair criticisms deriving from that. I am concerned that such issues may keep women away from academic research.

As a student, I didn't consciously look for role models when I was making decisions on what to study. Now, as a professor, I realize that the presence of women in the physics programs (both, among fellow students and mentors) is an important factor for young girls. I also realize that developing the right amount of self-confidence for academic life is a delicate process. As a matter of fact, you have to be self-critical to produce good science.

In summary?

I have taken a great next step in my career that factors in both my research interests, career development and enabled me to establish my family. This is really wonderful.

Prof. Rachel Grange (Swiss)

http://www.ong.ethz.ch
Optical Nanomaterial Group, Institute of Quantum Electronics, Department of Physics, ETH Zürich

Rachel Grange, "ETH Zurich/Giulia Marthaler"

Biography

Rachel Grange is (since 2015) an assistant professor of photonics in the Department of Physics, ETH Zürich. She completed a diploma in physics engineering at EPFL in 2002 and a PhD in physics at ETH Zürich in 2006. She worked as a postdoctoral fellow in Demetri Psaltis's group in EPFL, from 2007 to 2010. Then, from 2011 to 2014, she was a junior research group leader at the Abbe Center of Photonics (ACP) and the Institute of Applied Physics, Friedrich-Schiller University of Jena.

Research area

Her research focuses on material investigations at the nanoscale (1 nm being 1 billion times smaller than a meter). Her team develops high resolution imaging tools to study the photonic properties of unusual structures. Indeed, this field is dominated by metals and semiconductors for their reliable optoelectronic properties. In contrast, her laboratory investigates a new family of nanocrystals, the metal oxides, with more functionalities. One of her main goal is to find strategies to control and enhance the optical nonlinearities of materials at small scale to obtain versatile compact photonics devices.

Honors and awards

2016 ERC Starting Grant

Conversation with Rachel Grange, July 21, 2015

An academic position with research independence and also time with my three children—it has meant a less direct route to a professorship, but that does not bother me

How did you become interested in physics?

I only became interested in physics toward the end of high school. I come from a mountain region in Switzerland called Valais, which is in the French speaking region. I realized at 14 years old I wanted to go the academic route, to gymnasium, though this was unusual in our class; only two or three out of 17 went that way. Neither of my parents went to university, and instead went through the apprenticeship system, with my father setting up his own company to sell building materials and my mother, after she had children, staying at home.

I chose to study Latin and English as my specialism in gymnasium, it could have been Latin and Greek, but a modern language was more practical. After 3 years, we took all subjects, and my interest in math and physics grew and I took extra classes in the last 2 years. Until then my favorite subjects had been history, Latin and ancient civilizations. Once I realized that I wanted to go to university it felt important to do something that would help me get a job and live independently. I was really good at math, but that was too theoretical. Physics was more applied, and I had to work hard at it, you could study how things worked, and it had wider applications. I wanted to do something challenging and interesting, and physics was a topic where I needed to go to university to learn more.

What was your next step then?

I decided to do a Diploma in Physics Engineering at EPFL. At first it was a struggle to understand, but I was motivated to solve problems and enjoyed the learning process. The first 2 years we studied theory with math and physics, then it was more engineering and micro-engineering. We had 15% women in our class, so 10–15 females. A group of 3–5 of us worked together regularly, including two women. I still have good friends from that time. I also did a 2-month internship in a start-up micro-engineering company that made wheels for the inside of watches. It went on to be very successful and was good work experience. Then, when I finished my diploma in 1998, it was just before the Telecom Crisis of 2001 and companies were cutting back. Suddenly finding a job as a physicist was difficult, but I also felt that I needed more training, and would like to go into physics in more depth, so decided to do a PhD.

Were you already dealing with a dual-career situation at this point?

Yes, I met my husband at the end of high school, we are both from the same region. He is a medical doctor and he was working on his specialism in internal medicine and gastroenterology, both in Fribourg and Lausanne. His training involved 6 years of basic medicine, then 5 years internal medicine and 3 years gastroenterology. He needed to build up this experience. I did not want to go abroad for my PhD, and a tutor at EPFL recommended the group of Ursula Keller at ETH. My successful application enabled me to remain near my husband who was in Lausanne, to study physics further and gain fluency in High German.

Another reason I chose the Keller group was that I'd done my master's thesis on optics. I decided to focus on semiconductors, because they really interested me and I was researching thin films. It was a great experience in a good group, a big group, with 30 people, and my main mentor was a senior postdoc who was very good. We worked very hard, had lots of fun, and my husband came to Zürich for 3 years, though initially he faced language issues in the hospitals, because our mother tongue is French.

How, then, did you take the next steps in your working life?

At this point I did not have a mentor because my postdoc supervisor had moved to industry. I was a bit exhausted with the science and was fed up of working 11 hours a day in a lab. I decided to do something else—go into industry or become a patent lawyer. In the end, I got a permanent job in the Swiss State Secretariat for Education and Research (SERI). It was a counselor position, where I was the Swiss Delegate to the OECD on nanotechnology and designing new Swiss policies for that industry. I did data analysis, built the statistics for politicians and public, and studied bibliometrics. I am the type of person who plans well in advance, because I want to support myself financially so I'd found this position before the end of my PhD, and started immediately. However, the position made me feel too passive. I worked there for 1.6 years, but I began to realize that this was not a job I could do forever, even though it was a permanent position.

I needed to do more research and began to explore which professors at EPFL were working in Optics. By chance, in 2007, a United States professor Demetri Psaltis, an expert in nonlinear optics, had just arrived at EPFL from Caltech, US. He became the Dean of Engineering and had open positions. I moved to research using femtosecond lasers to study nanotechnology. This postdoc position was a great job, though initially I wondered why he selected me, but I had a lot of publications from my PhD time. I had no commuting (my husband's new position was in Lausanne) and I was working with a professor who had 25 years' experience in the United States and was establishing a group in Europe. It was great. We built up the lab from Caltech and there were only 10 people in the group, so we talked about science at group meetings. One postdoc had a lot of experience in Chemistry, so we were doing research with materials and learning new methods. We could change the setup every few days and could have a flexible approach to experiments.

It was around this time that you started your family, wasn't it?

Actually, my husband wanted us to start the family at the end of my PhD, so earlier than me. I tried to start a family when I worked at the government position in Bern, but I had a miscarriage. If I'd known in advance that this occurs in 25% of cases it would have saved me a lot pain. However, before I started at EPFL and after I had signed the contract, I became pregnant. It was awkward having to tell my new supervisor, but I was very confident that I could set up the lab before the baby was born. I started in September and the baby arrived 1 month early in January 2008. My son was almost born in the lab. I was going to go there on a Sunday, but I did not feel right so stayed at home. He was born the next day. I could not find public daycare in Lausanne, so we found a private place. I spent 2 months at home, then I returned to work at 50% for 4 months, working from 1 to 6 p.m. After 6 months, I was working 80%. Demetri Psaltis was flexible as long as I did my job, which helped enormously.

At this point, however, my husband moved to France for a year. He had to be flexible in order to find his gastroenterology fellowships, which meant moving to Normandy. I had only done 1 year of my postdoc and did not want to move. We managed this time by meeting every 3 weeks in Paris, which was possible because of the set up at EPFL. Working with my research team was great, we had a lovely team spirit. The PhD student was brilliant and did quality work in the lab, the colleagues were very nice and on top of this I organized social activities for the group. After a year, my husband returned. At this point I'd got a childcare place at EPFL, I was working 80% and it was easier and a really good environment. My second son arrived in April 2010, and then, in December 2010, I moved to the University of Jena as a research group leader. In all I did a 3-year postdoc in the Psaltis group and it was an excellent experience.

You always prepare and organize far in advance. What did you do to get the Jena position?

Yes, Demetri Psaltis said I could stay with him longer and was very supportive, but I knew I needed to take the next steps in my career. I spent 6 months looking for another job. I applied to many positions and was invited for interviews, and used the opportunity to practice through the whole process. I was invited to apply in the Netherlands, I applied for a position at King's College, London, in Edinburgh and in Jena. It was not the easiest time because I was breast feeding my son, so it was very draining.

After each interview, I rang them to get feedback and learn what my weaknesses and my strengths had been. This is an advantage that women can take, you have more opportunities to be interviewed, because of initiatives to increase gender balance. You can really learn from the procedure if you use it actively. I also talked with two to three assistant professors (AP) and asked them what they did to get their position. One key piece of advice was to prepare a budget and do it really well and the second was to be very confident during the interview, showing no sign of weakness or insecurity. You have to dare to ask these people questions, APs are on the next step and they remember really well what they had to do to succeed.

Another factor in the decision was my husband's career. He still needed to complete 8 months of his gastroenterology training. This meant we needed to choose a place where the language was English, French, or German. I was interviewed successfully in July 2010 for Jena and then my husband found a position in Jena too.

You and your husband have had to work actively to find compromises to make both of your careers proceed.

Yes, with a dual career often one of the people has to suffer, and it is hard to make compromises. When I chose my first postdoc it would not have been possible to go to the United States, so I had to compromise. Often those who go the United States come back to Europe and get an assistant professor position more easily. This is the traditional route. As an experienced postdoc, I wanted to find an academic position with independence. We more or less agreed that our goal would be Germany. At the University of Jena, I had a research group leader position, so this was the independent step. Somehow, we managed the complexity, though it was not easy for my husband in his job, because of the hierarchical structures in German hospitals.

How would you describe your experience at the University of Jena?

I had a fellowship from the Carl Zeiss Foundation, established by Jena since 1888. I was a group leader for 4 years. It was a really good experience. A further factor is that in Germany the family life provision was very good. I could organize daycare easily and the daily working hours were lower. It is an 8 a.m. to 5 p.m. culture. I always wanted to have three children so I took the opportunity in Jena to have another child. My daughter was born in 2012. In Germany, the parental leave is better and it was possible for both myself and my husband to share the childcare together. Indeed, for the first year it is just not possible to put the child into daycare. We managed the situation for 6–8 months with me working in the afternoons and my husband the mornings. I only took 2 months maternity leave, and went back 50% to enable this to happen.

When and how did you start to look for a new position?

There were a few open positions in Jena, but it was clear that the next step for my husband was to set up his own medical practice in Switzerland, which meant we planned to return to Switzerland. I applied for an SNSF professorship half way through my fellowship, as it takes one to 2 years for a successful application. My husband set up his practice in the Jura region, close to Basel. I'd had a successful time at Jena, and was very happy to get the SNSF professorship associated with ETH Zürich.

I am aware that I am older, that I have taken more time with my career and made a less direct route to becoming a professor, but this does not bother me. My goal is to get a full professorship and a permanent position, but I will not fall apart if it does not happen. It is important to be part of the community within my department. I also built up my research team in advance of my move to ETH, so I could start my research quickly. I was delighted that in 2016 I was awarded an ERC starting grant, which brought more funding, prestige and I could expand my group. I commute to work 4 days a week and spend Wednesday's at home with my family doing home office in the morning when the kids are at school.

What would you say are key factors for you in making the academic career work?

I have a very supportive husband and partner. We are a real team and very organized at home. I have achieved what I wanted and did not encounter huge obstacles. I am very lucky to have made it, but I know that I also did this myself. You have to take on board all your choices and take responsibility for them. I love teaching and think it is a wonderful part of a professor's life.

Any challenges in the community?

There are sometimes attitudes toward women that can be discouraging and could maybe kill ambition, though I was not affected by this except perhaps being told at EPFL by male classmates that females should not be in an engineering school. I try to say something now if I hear any of these kinds of comments, for the sake of young women researchers. I also experience that, as a female professor, more women come to my group and classes.

I think that the demands on scientists for mobility experience should be more flexible. I do not agree that people should be expected to go to the United States for 2 years. This is an impossible goal if you have a family. You should be able to count all your collaborations and shorter visits as mobility. We need excellence in research, but age limits and mobility criteria do not need to be so tightly defined.

Prof. Stefanie Hellweg (German)

http://www.esd.ifu.ethz.ch

Ecological Systems Design, Institute of Environmental Engineering, Department of Civil, Environmental and Geomatic Engineering, ETH Zürich

Stefanie Hellweg, "ETH Zurich/Giulia Marthaler"

Biography

Stefanie Hellweg has been full professor for ecological systems design at the Institute of Environmental Engineering since 2006. She was born in Essen (Germany) in 1971 and studied industrial engineering at Karlsruhe University, graduating in 1996. She worked as a management consultant on projects in the field of waste management and energy production. From 1997 to 2000 she completed her PhD at ETH Zurich. Between 2001 and 2002 she was a postdoc working jointly for Paul Scherrer Institute (PSI) and ETH on sustainable waste management. In 2003 she became a senior research fellow at the Institute for Chemical and Bioengineering at ETH Zurich and led several projects on the environmental assessment of chemical products and processes, the treatment of waste and wastewaters from the chemical industry, and chemical exposure at the workplace. She was a visiting scientist at Lawrence Berkeley National Laboratory (2004–05) and at Yale University, United States in 2013.

Research area

Stefanie Hellweg works on modeling, evaluating, and improving the environmental impact of products, technologies, and consumption patterns. In particular, she is developing and applying methods for life-cycle assessment and industrial ecology. Besides her research activities, she serves on several academic steering and advisory committees as well as editorial boards. She is member of the National Research Council of the Swiss National Science Foundation (SNF) and of the International Resource Panel of UN Environment.

Honors and awards

SNF research council member since 2011
Several personal SNSF career grants

Conversation with Stefanie Hellweg, October 30, 2015

When did it all start for you?

There was never any question in mind that I would not go to university. I really enjoyed high school and specialized in the math/english/physics Abitur. We also studied many other subjects but natural science was my focus.

I studied Industrial Engineering at Karlsruhe University, Germany. I chose this because it allowed me a broad view of science subjects (engineering, natural sciences and economics) and I had many interests. After receiving my diploma, I became a consultant and did not consider studying for a PhD at first. From my engineering program, it was a natural step to move into a more applied job. The consultancy was focused on energy and the environment, which was a good topic for me. However, working life was not how I expected it to be. Our goals were to deliver quick solutions, rather than undertaking in-depth study of an issue. I did not find this satisfactory or fulfilling.

My next goal was to do a PhD. I saw an interesting opening in my area and topic of interest. It involved combining environmental assessment with examining thermal waste treatment processes. The work relates to environmental engineering, but at that time the position was in the Chemistry Department at ETH Zürich. This was the beginning of my current research topic.

How would you describe your PhD experience?

I absolutely loved my PhD time and it was great to work on this topic. I had no thought to remain in academia, I just liked the subject. It did give me opportunities to take the next steps, with further interesting projects and ways to continue to work in this area.

During my PhD period, I had my first child and then my second arrived during the postdoc period. I was combining the research and having a family. My postdoc position remained in the Chemistry Department, but the work was done jointly with the Paul Scherrer Institute (PSI). Next, I worked for 3 years as senior research assistant in the Institute for Chemical and Bioengineering. During my time as postdoc and senior fellow I worked on a 60% contract.

How did you manage the research career and family life?

When I had my children, it was extremely difficult to find childcare. It was only after 9 months that I found someone to look after my child for 2 days a week. She was a nanny and, though at the beginning it was only for 2 days, it grew to full-time over the years. We have had the same nanny through the whole time with our children. Our nanny is really a good caretaker, perhaps it was better in the end that I could not find childcare places, because we found her. However, it would have been much easier to have access to a kindergarten earlier. My family and my husband's family do not live close, so we had to organize everything on our own. The result was I had my job and my family and I could not do anything more. There could be no distractions.

I was really helped by my PhD supervisor who was very flexible and absolutely great when my first baby arrived. He was child-friendly, flexible about when I worked, and I was able to bring the baby into work with me. In fact, I changed offices to share with a part-time employee so I could bring in the baby without disturbing colleagues. When my second child was born, our nanny had more time and was able to look after both children. Initially this was at 60%, but with time this grew. After the children went to school the situation improved, but the school system in Switzerland is geared to mothers who stay at home meaning that there were still problems. Nowadays this has improved with more facilities for children to do extra activities at school.

Some of my current PhD students find places at the ETH kindergarten, but many also resort to other childcare. It is much better than at my time, but getting a good childcare place remains a difficult task.

Your next step was to become a professor. How did this happen?

An associate professorship, meaning a tenured position, was advertised at ETH in my field and I decided to apply. I did not expect to be shortlisted, but I decided to try anyway, because I had nothing to lose. I was on sabbatical leave with my family at Lawrence Berkeley National Laboratory in the United States when the post was advertised, and I made the application from there and returned to Switzerland for the interviews. In the end, it all worked out and my profile fitted this new position, which was in the Department of Civil, Environmental and Geomatic Engineering. It was the ideal solution for me family wise. I had two children and could not move from Zürich. The timing was perfect, because I could start again working at 100%.

I live with a dual-career situation. My husband is also a professor at ETH Zürich, in the Computer Science Department. As we are both professors we understand the academic career and find ways to handle unforeseen events. We have had two sabbaticals taking the whole family. Sometimes the destination is a compromise. Berkeley was good for us both, but Yale in 2013 was better for me. However, my husband was able to travel, visiting people in his field.

How was your experience of becoming a professor?

It was exciting to achieve the associate professorship. I could establish my own group and then grow it with successful grant applications. There were a lot of teaching responsibilities, but I had an excellent group and so started the research quickly. In 2010 I became a full professor. It is very important that you are really interested in what you are doing. You need a lot of motivation, because it involves a big workload, for a long time, so it needs to be compelling.

Then in 2011 I was nominated to the Research Council of the Swiss National Science Foundation, which really gives one insight into the research landscape. My division is "Programs," dealing with national research programs, and therefore interdisciplinary research that has an impact on policies. This combination of research combined with applications for society is very interesting and also popular with students. The environmental engineering course also attracts higher rates of young women at 30%, which is really great for engineering.

What do you think influenced the progress of your career?

I was deeply interested in the topic. I was also lucky that opportunities turned up at the right time, which made it easier to follow the path. However, I was ready to apply for positions, even if they seemed impossible.

Were there any challenges with your dual-career situation?

My husband got his professorship first, which meant I made the first compromises. When I had no childcare at home initially we had to find a fair solution that worked for us both. In the early stages I was more with the children, but when I got my professorship this became more balanced. The childcare situation when my children arrived was really bad, and the biggest help was my supervisor's flexibility. Now there are many more kindergartens, and support for female academics with babies.

What do you need in order to be a successful professor?

You need to be good at a number of skills: generating new ideas and developing research areas, while managing the teaching load. You need to find good researchers and people for your group, to choose a good research topic and place those results within your scientific community. On top of this are the management roles, which professors are not trained for. You have to handle budgets and be responsible for personnel matters in your group.

How would you describe your positive experiences?

I did not have specific mentors in my career, but the people I worked with were understanding. I always felt very welcomed and well received, in constructive and friendly environments. There is the tough process of fighting for resources, but that is normal. A big motivation is that I am paid to do research that I really like and is worthwhile. As professors, we can shape our research directions and future careers.

What would be your advice to young women?

Try to organize yourself really well if you have a family. You need to be as good as other researchers, if not better. Do not believe that there are any bonuses because you are a woman, because in reality that is rarely the case in academia.

You have significant experiences sitting on commissions?

I have experienced some situations where women were sabotaged or heavily criticized. This can be particularly bad if it comes from a hiring commission and if it is unconscious. There is that usual habit of calling confident women "boasting," while the same behavior with men may be positively viewed. As a member of a commission I try to mention the contradictions and turn these attitudes round. There are many requests to sit on commissions, particularly for women, so I think it is also important to sometimes say no and manage your time well.

What would be your advice for women who are taking the steps in the academic career?

As for anyone, the most important factor is to conduct convincing research. To be confident really helps you, but there may be a problem when women are aggressive, it may frighten and annoy people. It is also important that you are ambitious enough to take the next step and you don't stop too early. Keep going and don't limit your possibilities. For example, it might be possible to go abroad for a shorter time, say 6 months, and the whole family can profit from a time abroad. You need to have a partner that works with you on this. Every person will find their way but don't set your goals too low!

Prof. Ursula Keller (Swiss)

www.ulp.ethz.ch
Ultrafast Laser Physics, Institute of Quantum Electronics, Department of Physics, ETH Zürich

Ursula Keller, "ETH Zürich/Heidi Hostettler"

Biography

Ursula Keller has been a tenured professor of physics at ETH Zurich since 1993, and serves as a director of the Swiss research program NCCR MUST in ultrafast science since 2010 (www.nccr-must.ch). She received a "Diplom" at ETH Zurich in 1984, a PhD at Stanford University United States in 1989, was a member of technical staff at Bell Labs United States 1989–93. She has been a cofounder and board member for Time-Bandwidth Products (acquired by JDSU in 2014) and for GigaTera (acquired by Time-Bandwidth in 2003). Her research interests are exploring and pushing the frontiers in ultrafast science and technology. Awards include the ERC advanced grants (2012 and 2018), European Inventor Award for lifetime achievement (2018), IEEE Photonics Award (2018), Weizmann Women and Science Award (2017), OSA Charles H. Townes Award (2015), LIA Arthur L. Schawlow Award (2013), EPS Senior Prize (2011), OSA Fraunhofer/Burley Prize (2008), Leibinger Innovation Prize (2004), and Zeiss Research Award (1998). In 2012 she founded the ETH Women Professors Forum and served as elected president from 2012 to 2016. (www.eth-wpf.ch)

Research interests

Prof. Keller's research interests are exploring and pushing the frontiers in ultrafast science and technology: ultrafast solid-state and semiconductor lasers, ultrashort pulse generation in the one to two optical cycle regime, frequency comb generation and stabilization, reliable and functional instrumentation for extreme ultraviolet

(EUV) to X-ray generation, attosecond experiments using high harmonic generation, and attosecond science.

Recent honors and awards

2018	European Inventor Lifetime Achievement Award
2018	IEEE Photonics Award
2012, 2018	ERC Advance Grant
2017	Weizmann Women and Science Award (full list see http://www.ulp.ethz.ch/people/kursula/honors-awards.html)

Conversation with Ursula Keller, March 29, 2016

"Take charge of your life. You have the power to make good things happen!"

When did you first become interested in science?

The turning point for me was when my math abilities were recognized at about 16 years old. Until then I had taken the nonacademic route, which was the tradition in my family, but my exceptional math ability enabled me to change routes and move to the gymnasium. From that point I aimed for university.

People found it strange that I chose to study math or physics at university, and my father needed to be convinced too. He understood the choice of math but thought physics too technical. He financed my studies and he wanted me to go to the best place possible (i.e., ETH Zürich) and to prove that I could be a success. From an early age I always wanted to study subjects that were related to finding a good job and to become financially independent. Maybe this influenced my choices at this point too.

What was your experience of being at university?

My main aim at the beginning of my ETH studies was to develop a good social life and it was easy with other physics students, because we could relate to each other. I went ski mountaineering every weekend with students in my year and I loved it. It helped that I was a good skier. I was very confident in my abilities, because at the gymnasium I did not have to study hard to do well in math and physics. I just understood it all, could do it, and even earned money for teaching math to younger students. In the first year at ETH Zürich I partied and still got exceptional results, in the second year the marks were still above average, but I did not invest enough study time to get the highest marks.

You went to Stanford University for your PhD, on a Fulbright Fellowship, what made you think of studying abroad rather than remaining in Switzerland?

I decided quickly that I wanted to have experience outside of Switzerland, to travel and study at the same time was a great option. At the beginning of my second year, I checked out all possible scholarships for doctoral study in the United States, and I decided to aim for a Fulbright. I needed to achieve top marks in my exams, and in years 3–5 of my diploma I invested more time in my studies. I guess that is part of my character: I think early about future paths, invest in finding out information, and keep open all possibilities.

In the end I got good grades, in fact I came top of my year, and I immediately applied for a Fulbright award. The process took a year to complete, which meant I needed to find constructive ways to spend that year. I asked an ETH Zürich professor if I could do a project in his lab, but he did not have a place available. Instead he suggested I go to the United Kingdom as an intern in a Physics lab and, at the same time, improve my English. He could put me in contact with a group at Imperial College, London or at Heriot-Watt University in Edinburgh. I chose Heriot-Watt, because I wanted to visit the Scottish Highlands. It was one of those moments when things happen fast: the professor phoned the Head of the Physics Department in Edinburgh immediately, I was offered an internship straight away and accepted on the spot. It was vital to be open to an unexpected opportunity, even though I had a boyfriend at the time.

What were the outcomes of the experience in Heriot-Watt?

At Heriot-Watt they studied Optical Bistability and Optical Computing. I joined a collection of excellent people and I had a great time socially. We did lots of hiking and skiing. After my year in Scotland I was fluent in English (albeit with a Scottish accent!) and was able to start university immediately in the United States.

You did your PhD at Stanford University. How did you choose this school?

For the Fulbright grant I applied to five schools: Stanford, Berkeley, Caltech and Arizona and Rochester (because of their good reputation in Optics). I was accepted at them all, and the Scottish research group advised me to go to Stanford, because of its international reputation. Another outcome from my stay in Scotland was the realization that, coming from the continent of Europe, I was behind age-wise in the graduation process. In Scotland people were finishing their PhDs at 25, while at 25, I had not yet begun mine. It made me resolve to speed up my PhD period once I arrived in the United States.

The Fulbright award financed my first year giving me time to focus on finding a research group, a supervisor and a research area. Geraldine Kenney-Wallace, a visiting professor from Canada, was very influential for me in that first year. She took me into her directed study program and understood that I wanted to finish my PhD quickly. In fact, she found the right group for me and spoke directly with the professor, Dave Bloom, who was a new young professor. I changed research groups to join his team; he was dynamic, result-orientated and I was only the fifth student in his group.

Another significant experience that year came from attending a recruitment reception organized by AT&T Bell Labs. Being Swiss, I arrived 15 minutes early, and had the opportunity to talk on my own to the recruiters. I even asked if it was possible to apply for a Bell Fellowship to sponsor the rest of my PhD, but only United States citizens were eligible. However, as a result of this meeting Dave Auston (a Canadian working at Bell Labs) offered me a 3-month job in New Jersey for the following summer. I drove my car across the United States and back, which was quite an adventure.

Bell Labs then offered me a PhD position, based in New Jersey, while maintaining my links to Stanford. It was a great opportunity and I weighed up the options—but in the end Stanford was the best place for me. I had the possibility to finish my PhD in about 3 years, if all went well with the research, and I had the sense that there were many opportunities there.

Can you speak more about your PhD experience?

I was interested in many topics during my PhD, ultrafast processes for example, and I was always seeking out new areas of research. Initially, older PhD students and postdocs were my mentors, and then I developed more independence. I started my interest in pump-probe work there. Dave Bloom did not micromanage, which made me very independent. I also developed my collaborations with Bell Labs further and began to get visibility in the field. One-time Dave Bloom sent me to a conference to present one of his invited talks. The organizers were initially upset about the substitution and it was not clear, until the last minute, whether they would allow me to talk. There was such a fuss that, by the end of the conference, everybody knew who I was, especially because my presentation went very well.

During my PhD I was part of a good student network, which supported my job search process. I went to a lot of interviews and was offered permanent positions at IBM, Bell Labs and other places in Silicon Valley. My dream job offer was from Bell Labs in New Jersey. At this point I was not thinking of becoming a professor.

You lived in California and your dream job was in New Jersey. Did you face a dual-career situation?

Yes! My husband worked in a start-up company in Silicon Valley, CA, and my dream job was in New Jersey thousands of miles away. My husband supported my choice to take the position with Bell Labs, though initially he was not so happy about the idea. If you are married and don't have kids, you don't have to live in the same place. The distance made our relationship stronger, we talked everyday, and when we met our meetings were romantic. In between we both worked hard at our jobs and research. I was at Bell Labs for nearly 4 years and during that time I did wonder how we would find the place where we could both work and live together.

The position at Bell Labs was prestigious, with the opportunity to do international level research, which meant that after 3 years I was approached by United States universities to see if I might be interested in a professorship. I was offered professorships at the University of Berkeley and the University of Michigan. I had also maintained contact with ETH Zürich professors, PhD students and postdocs when I visited Switzerland annually to see my family. Then, when the offer from ETH Zürich came, I was a perfect fit; the work at Bell Labs had enabled me to take the path of excellence with my research results. Continually striving for research excellence opens doors, but it's important to maintain contacts too. I negotiated with ETH Zürich to start my professorship with tenure: my job at Bell Labs was permanent, my husband was moving with me, leaving his job in California, and intended to establish his own company. I negotiated hard for my start-up funds.

In 1993 you became a tenured physics professor at the age of 33 years. How would you describe your experience?

I underestimated what it would be like to be the first tenured female physics professor. I had a very good international network of colleagues in Bell Labs and in the photonics network in Europe. However, without realizing it, I was not integrated into my department at ETH. The recognition that there was a problem came only when I got pregnant (my sons were born in 1997 and 1999). My colleagues' response was not positive and they delayed my promotion to full professorship. It took 3 years for my male colleagues and 4.5 years for me.

It is only in the last few years that I feel integrated into Switzerland. This came as a result of becoming the Director of NCCR MUST (2010) and with the creation of the ETH Women Professors Forum (ETH WPF) in 2012. Until then I had felt pretty isolated. My field of ultrafast science was not represented at ETH, so I was orientated toward my international academic peers. I was focused on my family and more distant from the local physics community than I realized. I concentrated on my research field and worked hard. However, one advantage of being less integrated was that I did not receive invitations to sit on committees, so I was able to concentrate on research without distraction.

With the creation of the ETH WPF I acquired female colleagues and I learned new political skills. I learned a lot from other women at leadership level, such Prof. Janet Hering, the Director of Eawag, a large ETH Research Institute. As a result of the Directorship of NCCR MUST I was also invited to sit on the Research Council of the Swiss National Science Foundation and became part of the group that assesses funding proposals from Swiss Universities. This means that I am embedded in a broader swiss-wide research network.

Could you identify any significant turning points in your career?

The events that lead to me becoming the Director of NCCR MUST in 2010 were very significant. It arose from a situation where my initial application for an important research grant—the ERC Advance Grant—was not successful. It was a shock and I ended up ruining a week of vacation feeling bad and sorry for myself. However, at the same time, another large joint Swiss research application came to fruition and I was invited to become the Director of a Swiss-wide research network working on ultrafast science in physics and chemistry. I would not have considered this position had my ERC grant been approved. The Directorship of NCCR MUST opened up my world in terms of leadership, women scientist networks and gender issues. I applied for an ERC Advance Grant for the second time in 2012 and was successful.

I have always stayed open to new possibilities, but this experience showed me how, what appeared to be a significant set-back, actually enabled unexpected opportunities to open up, which then resulted in expanded, influential and fulfilling responsibilities.

What factors would you say resulted in you becoming a professor?

Well, as you can see, I did not have a carefully planned trajectory toward academia. I was focused on doing the most exciting science projects and publishing at the top level, and I kept my options open. Though my CV looks well managed, I was simply taking advantage of the best, and most interesting, opportunities that came my way, leaving the next steps open. By the time I was approached by the universities I was looking for a new challenge and was excited about the potential of moving to academia.

What advice would you give to others thinking of becoming a professor?

I recommend that you spend time to understand yourself and what drives you. Identify what it is you enjoy doing and what might be key factors in your choices. For example, I always wanted to be financially independent, which meant I had an early emphasis on a career. Further, it's important to understand your attitudes to mobility and work. I would recommend that you don't let this limit your options. It also helps if you ask yourself early—what are my next potential ways forward? I was always thinking far in advance. It enabled me to stay flexible about choices, and this expanded my possibilities. I think if you understand yourself and can answer these questions it reduces stress. Each person has a different career path, does things differently and works in different environments. There is no formula for success, but you can find your way.

Are there any other recommendations?

The main factors which helped me are as follows:

- you should strive to deliver the best quality in your work;
- set up collaborations with the best possible people;
- think in advance about all options and target good places to work;
- always make space for free time activities (a driving factor for me was to do sport, have a good social life and travel).

On working with a dual-career situation?

My husband and I are totally equal. His goal was to establish his own company and my professorship was a secure job that gave him time to start a company in Switzerland. As part of the dual-career negotiation he benefited from the newly created ETH spin-off program. He had a 50% job in my group for 2 years, was able to learn German, and then get established in Switzerland. The company Time-Bandwidth Products AG was so successful, that it was acquired by a major United States company in 2014.

From an early stage in our careers we separated our work responsibilities and space. I was in the academic world and he was in start-ups and business. We had a totally equal partnership with the children. We waited to have children and were prepared to go without them, then when I was 37 we decided to try. It worked smoothly, we had two boys, and having children broadened our horizons. Initially I underestimated the extra workload arising from the arrival of a new baby and it was quite a shock; I thought that maternity leave would give me time to do more work! I had very good pregnancies physically, so this was all a very positive experience, but of course our lives changed a great deal.

Would you say that there were particular challenges in your career?

My main challenge was my local isolation in the Physics Department at ETH Zürich. I was not initially aware of this, but it hurt on multiple levels. If an ETH Women Professors Forum had been established earlier, it could have helped a lot with this, because I was oblivious about how to deal with the issues in my local community. When a generation change came in the Physics Department I did not pay attention to cultivating relationships with the new generation. I did not do enough local networking and this came at a high price later. Women are often not networked enough and need to make a special effort. By the time I reached out, it was difficult to build relationships. In the end my colleagues see me only as a competitor for the limited resources in the department. Luckily, I always concentrated on achieving scientific excellence, and this drives me forward.

The second challenge came during the time we had small children (and this can apply equally for fathers involved in parenting), because it is also easy to lose contact then with your local community. At this point you need a sponsor who keeps you engaged, such as a friendly colleague who can tell you which events are important to attend. For women it is physically harder during the small child phase, which, if you have two children, can last from 4 to 5 years. A Women Professors Forum can support at this time; you can call female colleagues for advice. In my opinion women can give each other the best advice—for example women professors don't assume that you will be satisfied with half, or less resources, just because you are a mother. They can also give advice on when to take leadership roles, this is important as you need to be selective about taking on committee work. Often, given that women are in a minority, they can receive many invitations and become overcommitted, to the detriment of their research work.

How would you sum up your life as a professor and your career?

When I think about my life I consider that I have an incredible situation. I have a dream job, working on exciting research, with huge independence and significant resources. I have a great partner and family and we have wonderful lifestyle. Being a professor brings so many benefits to life, and these outweigh any challenges you might have deal with. I can only recommend it!

Prof. Salomé LeibundGut-Landmann (Swiss)

http://www.vetvir.uzh.ch/en/Research/Immunology
Institute of Virology, VetSuisse Faculty, University of Zürich

Salomé LeibundGut

Biography

Since April 2017 Salomé LeibundGut has been an associate professor, Vetsuisse Faculty, University of Zürich. Between 2010 and 2015 she was an SNSF assistant professor, Department of Biology, ETH Zürich and from 2015 to 2017 at the VetSuisse Faculty, UZH. She studied a diploma in biology at ETH Zurich graduating in 1998. She obtained her PhD in biology from the University of Geneva in 2003. From 2003 to 2004 she was a postdoctoral fellow with Prof. Walter Reith at the Institute of Pathology and Immunology, University of Geneva. From 2004 to 2008 she was a postdoc in the lab of Dr. Caetano Reis e Sousa at Cancer Research UK, London, United Kingdom. She worked as a senior scientist, from 2008 to 2009, in the lab of Prof. Annette Oxenius, Institute for Microbiology, ETH Zurich.

Research interests

The Immunology section investigates innate and adaptive defense strategies against human and animal pathogens with a special interest in fungal pathogens. Understanding the basic mechanisms of fungal pathogenicity and antifungal defense is key for improving diagnostic, therapeutic and preventive measures against these clinically relevant opportunistic infections. The LeibundGut laboratory focuses on immune mechanisms against *Candida albicans* and *Malassezia* spp., which protect the host in a tissue specific manner from mucosal and systemic infections. A particular interest is in the role of interleukin-17 and neutrophil-mediated antifungal defense. The knowledge gained from this research is also relevant for gaining a better understanding of host defense against other infectious agents and tumors, as well as gaining insights into the crosstalk between infectious and inflammatory disorders.

Honors and awards

2014 Hedi Fritz-Niggli guest professor, University of Zürich
2013 ETH President Award for Outstanding Achievements
2009 SNSF Professorship

Conversation with Salomé LeibundGut-Landmann, June 23, 2015

A series of important mentors helped me to develop a step-by-step decision-making process, which was vital on my path to professorship.

Where did the interest in science start for you?

I was always interested in biology. As a child, I spent hours in my grandfather's garden gardening. Later I became more interested in biomedicine. My father was a surgeon and my mother a medical doctor researching infectious diseases. One of my sisters went on to be a physician too. After seeing my family experience, being a doctor and caring for patients was not for me, but instead I wanted to study biological processes in research. I chose to leave home after high school and start an independent life studying a Diploma (equivalent of nowadays MSc degree) in Biochemistry and Molecular Biology at ETH Zürich.

Before I went to ETH, I undertook an internship at the Scripps Research Institute in La Jolla, California. I had the chance to do lab work (and improve my English skills) for 3 months, and this experience reinforced my choice of subject. I was lucky to have a great mentor in the lab who explained everything to me, despite my very limited knowledge at that time. I helped PhD students with their work, so I had the opportunity to see what it meant to be a working scientist before I even started university. This really helped when I went through basic lectures during the first years at ETH, because I understood how the topics would be useful in the workplace later.

I recommend doing internships early during a career. In fact, I continued this type of experience throughout my undergraduate studies. For example, I worked for several weeks in an allergy research institute in Davos, which gave me the chance to see another research environment and I could contrast the way that labs are run. For my Diploma thesis (a master's research project) I was based in the lab of Prof. Tim Richmond who worked on X-ray crystallography. I was given a solid knowledge in molecular biology and the techniques used for generating recombinant proteins that were later used for crystallization, but I became aware that structural biology was not the career I would pursue. So, I said no to the offer of a PhD position there.

Why did you choose to your PhD at the University of Geneva?

After my Diploma, I did another 3-month's internship, this time in the lab of Prof. Rolf Zinkernagel, the 1996 Nobel Prize laureate for Physiology and Medicine. He won the prize for research that discovered MHC restriction, a principle that defines the specificity of the cell-mediated immune defense. I assisted one of his postdocs, found the research fascinating and realized that immunology was the topic for me.

This internship gave me important insights in the field. I had always assumed that I would do a PhD and I always wanted to continue with research. I never questioned this path, because it just felt right. I had observed my mother's experience and it was always clear I wanted this route. I did stay open-minded to alternatives, but those that presented themselves were never as attractive, so I did not change direction.

I applied for a PhD position in the Zinkernagel lab. It felt like a tragedy at first when I did not get the position. I then checked out options at other immunology labs throughout Switzerland and was attracted most to research in the field of molecular immunology at the University Geneva. I stayed for 6 years in the laboratory of Prof. Walter Reith, studying the transcriptional regulation of antigen presenting molecules in dendritic cells. It was a productive and also very pleasant time. However, at some point I realized that I needed to move on for my career.

Were you looking for research experience abroad?

I don't think it was as deliberate as that. I was not actively looking for a position. I visited the lab of Prof. Caetano Reis e Sousa at Cancer Research UK (now The Francis Crick Institute), in London, for a week's research collaboration to study the function of dendritic cells in immunity. This experience made me want to join his lab for a postdoctoral fellowship. I applied successfully for an SNSF mobility grant and an EMBO long-term fellowship and went on to spend 3 years in London. My husband remained in Switzerland, which was not easy, but luckily, in the last year of my stay abroad he could join me by taking a sabbatical. We spent a great year exploring a fascinating metropole.

My time as a postdoc in London was marked by exciting science, a vibrant research environment and great colleagues, all of which was inspired by the personality of the lab head, who is not only an outstanding scientist but also an excellent mentor. After some initial struggles to find my niche in the group, I was lucky to work on the immune response to fungal pathogens and to discover a new function for a fungus-specific pattern recognition receptor in mammals. This increased my fascination for fungal immunity.

I recommend selecting a lab for the postdoc period not only by its research topic, but also by the quality of its PI. He/she can be an invaluable support when it comes to taking the next step(s).

What did you do after the London postdoc period and how did you get there?

I knew that my next step would be to move back to Switzerland, but I did not see clearly how to reach the next step toward an independent research career. I really had to deal with what they call "the imposter syndrome" feelings during this time. I was lucky enough to be referred to a mentoring program for young PIs in London. It had been set up by a female director of the Cancer Institute and I was accepted, although I was still a postdoc. My mentor helped me see my inner blocks and challenges, and helped me find possible solutions, which was very strengthening. Her central advice was that I take one step at a time, and then see what would happen at each step. I was aware that I might fail at each step, but this approach of not thinking too far ahead, kept me optimistic. I approached different people to get a sense of possible options. I was lucky to speak to Prof. Annette Oxenius, a professor from ETH Zürich, whom I knew from my internship in the Zinkernagel lab where she used to be a postdoc. She encouraged me to consider moving to ETH Zürich. Her institute was interested in my research topic and offered the space to house my potential group in case that I could bring my own funding. I thus applied for an SNSF professorship which, if successful, would bring 6 years of funding to establish my group.

How did you manage the process of change from London to Zürich?

It was a 2-year process to get the SNSF professorship and I came back to Switzerland before knowing the outcome. I joined Annette's lab for an additional postdoc, which enabled me to be integrated into the Institute of Microbiology before I received my award. I gave birth to my son during this time and had to handle the organization of the family life. It was a transition period between postdoc, new mum and establishing my own lab, which gave me time to adapt to all the personal and professional changes.

I'm the kind of person that needs time to develop, to absorb the process at each stage and to mature. I was able to learn from the labs around me and decide how I wanted my lab to be; I had time to equip the lab space and to recruit my staff in advance. On Day 1 of my SNSF professorship the lab was already running and we started experimenting. My lab was embedded in the bigger structure of the institute and we had joint lab meetings with Annette's group. This meant a great deal of support for our projects, while we complemented their research questions with our independent topics.

You were handling becoming a mother at the same time as the transition to professor?

Yes, and my second pregnancy came near the beginning of the SNSF professorship and I gave birth to my daughter after my lab had been running for 9 months. During the maternity leave I had regular meetings with the group to supervise their projects. I did not take a long break, but I enjoyed the extra time with the family. Having two children is more fun, though it means a lot of work. It helps that we have a wonderful nanny.

How would you outline your research area and particular community?

The focus of my research is on infection biology and on the immune mechanisms to fungal pathogens. Fungi can cause very serious diseases, especially in immunocompromised individuals. Fungal infections remain understudied despite their clinical relevance. This makes them an attractive area of research, as much remains to be discovered, and more research is clearly needed to meet the clinical needs. The collegiality in our research community is an asset as it makes collaborations fun and rewarding. I enjoy visiting scientific meetings, as it means meeting friends, talking science with peers and developing new research projects, while taking a break from the everyday responsibilities.

Were you still managing a dual-career situation at this time?

Yes. We both work full time, but we both benefit from flexible working hours. Thus, we share the childcare and the children benefit from two equal parents, each of us having his own style of parenting. It helps them grow into independent personalities.

One thing that has been positive for me is that we are in different fields professionally. My husband is in finance. He is very good at managing projects so we often discuss overall issues and management issues together. He is aware that my science career and lab are important to me and he rarely complains if I work long hours. We both respect and support each other's passion. It helps that we complement but don't compete each other.

The SNSF professorship is a temporary position by definition. What was your next step?

It was always clear that my assistant professor (AP) position at ETH was non tenure-track. I received several hints from senior colleagues, who had followed the same path, about the challenge. I was not put off by it though. I recognized the chance to start my independent career in a great environment and this was what counted most at the time. There were several APs at the Department of Biology in the same situation as me, without a stable position, and we set up a forum to discuss and support each other.

Toward the second half of my appointment, it got somewhat stressful to find the next step. I started to search out opportunities, and within a short time I was offered a tenured position at a University elsewhere in Switzerland. However, for family reasons I could not move from Zürich. As commuting was not an option either, I felt obliged to decline. I did not want to put stress on a stable family situation. This decision caused many sleepless nights. It was not obvious that I would find an alternative way forward.

You have just moved to the University of Zürich to become a member of the Vetsuisse faculty. How did this happen?

I got in contact with the faculty there, as they were looking for an immunologist; it was a gap in their specialism. My inquiry was timely as there were common interests. I moved while I had still my SNSF professorship, with my funds transferred to the University of Zürich, but the plan was always to obtain promotion to a tenured position.

You were a Hedi Fritz-Niggli guest professor at University of Zürich for 3 months before you joined the university? What did that involve?

This is a fellowship for visiting female professors whose job is to encourage young scientists in academia. The idea behind it is built on the principle of positive role models. It was a great opportunity for me, because I got to know the Vetsuisse faculty and many future colleagues and I was able to organize mentoring and workshops for young people through relating to everyone on the important topic of mentors. I had such good mentors in my career, who made a significant difference to me, that I want to give something back to younger scientists. I have experienced how at certain times ideas and suggestions from experienced academics can make a difference. I was helped to open up a door into research, to find the next step, to evolve, step through more doors, while at the same time checking if each step felt right.

My key advice to aspiring professors is that you should not wait for open positions to be advertised, but rather look around early. Approach departments or institutes, talk with people, discuss how your research may fit, search options and make visits whenever possible. Job offers often come from discussions with people. Everyone will have an individualized path, but you need to look out yourself for those hidden opportunities.

In 2017 you were promoted to associate professor? How has your position changed?

My current position comes with all the duties that a tenured position implies, such as teaching, commission work, and administrative burdens. But at the same time, holding an academic position is a great privilege, because it means interacting with inspiring people, promoting science, and pursuing my passion.

Prof. Ulrike Lohmann (German)

http://www.iac.ethz.ch/group/atmospheric-physics.html
Atmospheric Physics, Institute for Atmospheric and Climate Science, Department of Environmental Systems Science, ETH Zürich

Ulrike Lohmann, "ETH Zurich/Giulia Marthaler"

Biography

Ulrike Lohmann is full professor for experimental atmospheric physics in the Institute for Atmospheric and Climate Science since October 2004. She was born in 1966 in Berlin (Germany) and studied from 1988 to 1993 meteorology at the Universities of Mainz and Hamburg. In 1996 she obtained her PhD in climate modeling from the Max Planck Institute for Meteorology. She was a postdoctoral fellow at the Canadian Centre for Climate Modelling and Analysis in Victoria (1996–97) and assistant and associate professor at Dalhousie University in Halifax (Canada) (1997–2004).

Research area

Ulrike Lohmann's research focuses on the role of aerosol particles and clouds in the climate system. Her specific interests are cloud microphysical processes including the formation of cloud droplets and ice crystals and the influence of aerosol particles on the radiation balance and on the hydrological cycle in the present, past and future climate. She combines laboratory work and field measurements on cloud and aerosol microphysics with the representation of them in different numerical models.

Honors and awards

She was awarded a Canada Research Chair in 2002, received the AMS Henry G. Houghton Award in 2007, was elected as a fellow of the American Geophysical Union in 2008 and of the German National Academy of Sciences, Leopoldina in 2014. She was a lead author for the Fourth and Fifth Assessment Reports of the Intergovernmental Panel for Climate Change (IPCC). She was the coordinator of the EU FP7 project BACCHUS and chaired the ECHAM-HAMMOZ consortium. At ETH, she was the head of the Institute for Atmospheric and Climate Science from 2006 to 2014, is a delegate of the ETH president for heading search committees and vice president of the Lecturer's conference. In 2018 she became a member of the Research Council of the SNSF and was awarded a honorary doctorate from Stockholm University.

Conversation with Ulrike Lohmann, August 5, 2015

"Make the most of chance encounters, they may bring unexpected opportunities"

How did you become interested in science?

It was through my parents. Math was my first love and when I felt bored as a child I asked my father to give me math questions to solve. Even in a restaurant when I was waiting for the next course, I would do math. My father was a politician and my mother a high school teacher. So, I grew up in an academic environment. As a child, my toys included experimental things, such as building something with electrical components, or using a chemistry set. I was a bit of a tomboy, but I also wanted to have friendships with girls. Sometimes I found it boring if all the girls did was drawing, or playing with dolls, but they were still good friends.

How did you choose your degree topic for university?

I did not have a straight career path at all. At one point in my teenage years, I decided to really rebel and not to study. As a result my final marks for the Abitur (high school degree) were only 2.5, instead of the 1.0 needed for some university programs.

What saved me was my decision to go abroad to Nigeria for a year after high school with an International Christian Youth Exchange. This was in 1985, and I went to Lagos, a huge city. I traveled with six people, one of them from Germany. She was 4 years older, and I felt as though I had a big sister, which was nice. Apart from that, my experience was like being thrown into ice cold water. I knew that I had to change my attitude if I ever wanted to study and realized that to do this I had to move away from home. I hoped that the year would be life-changing and it was. It was really tough in Lagos, my job was only part-time, so I had to manage with a lot of free time. We had no running water or electricity, we had to shop everyday and there was often no bread. Then we had to cook three times daily. The contrasting life there put things into perspective, but I was still undecided on what to study on my return.

So how did you end up with a diploma in meteorology?

The whole decision-making process did not go smoothly. First, I studied ethnology, African studies, and sociology for a semester, which made sense after my stay in Africa, but it was not right for me, so I went to the student guidance office.

I would have liked to study environmental engineering in Berlin, but my high school marks were simply not good enough for me to get a place on that course. I was advised to study physical geography, which meant that I also had to take minor courses. By chance only, I did a minor in meteorology, which basically was an introduction to atmospheric science. This course was based on physics and I enjoyed it a lot. For the other minor I chose chemistry, but then I failed the chemistry lab course after 1 year. I could have redone it, but I thought that I should not be studying chemistry if I could barely pass the course. At this point I went back to my parents, who were financing me, and said "I want to study meteorology." We made a deal—either I succeeded in meteorology, or I had to finish my geography degree.

What did you have to do in order to make this next change?

First, I had to do what is called a Vordiplom—which was basically a 2-year course in math and physics. I worked really hard, studied like hell and enjoyed the courses a lot. My enthusiasm for this topic basically switched on overnight. I was not really sure that I was good enough, but I enjoyed every part of it, including solving problem sets and studying for the exams.

My work for the Vordiplom proved to me that I could do it. I got top marks and then was able to move to the University of Hamburg to study meteorology. When I finished my diploma, I moved to the Max Planck Institute for Meteorology, which was next to the university, to do my PhD.

In short, I had a 3-year long turning point after high school before I found out what I wanted to do and I only found my subject by coincidence. I had explored so many things, which meant that I was much more mature when I finally found meteorology as the topic of my choice and interest. The whole long route to finding this topic affected my self-confidence deeply and I was not convinced that I was good enough for a career in science. However, I really got to love climate science and was hoping that I could return to work at the Max Planck Institute as a senior scientist after some international experience.

You moved to Canada as a postdoc. How did this happen?

It was a mixture of factors, some experience and others coincidence. I knew after my year in Nigeria that I needed to work in a community where I felt more integrated. In Lagos, you were always a novelty and always visible and that was tough.

Once I had my PhD it was clear to me that I would want to do a postdoc. We had a visiting scientist from Canada at the Institute and he suggested that I go to Victoria, BC to Environment Canada, which had an institute for climate science there. As a result, I applied for a Canadian NSERC fellowship and got the award. I was based at the Canadian Climate Modeling Center in Victoria, but I actually did not really enjoy working there. It was a rather bureaucratic environment that I did not appreciate. Everything felt restrictive, because you could not travel without permission, even if you got invited to conferences and got your expenses covered.

Again, I questioned what I was doing and what I should do next. I did not enjoy carrying out climate modeling on the computer all by myself. I missed the lively atmosphere of the Max Planck Institute in Hamburg to which I would have loved to go back. However, at this time I attended an atmospheric science workshop in Canada and I met a professor from Dalhousie University in Nova Scotia, Canada. He approached me and asked if I might be interested in a tenure-track professorship at his university. I had never heard of the university and thus was at first sight not interested. But then I thought that I had nothing to lose by applying for this position, given that Dalhousie was an unknown university to me back then. I felt that there was less pressure for performing well. I got the job and, suddenly, I had a tenure-track professorship after 1 year of postdoctoral research.

This was the best thing that could have happened to me, where the opportunity came out of a chance encounter. Dalhousie had a small atmospheric science program with only four professors, two hosted in the Department of Physics and two in oceanography. I had the independence to do as I wanted and I was able to work on aerosol particles and clouds, and also to attract a lot of money for my research area. It was such a small group that I ended up starting to work experimentally myself.

What was it like to work at Dalhousie University?

I was one of the two atmospheric scientists in the Physics Department, but the physicists left me alone. Someone asked me how it was to be the only woman, but the other factors were that I was 20 years younger than the other faculty and not a "real physicist." Pure physicists thought that atmospheric work was "soft physics" so I was left by myself. During this time, I traveled extensively developing connections outside as much as possible. I had a visiting professorship at Columbia University, New York and did many exchanges with them. I was a Dalhousie University for 7 years and was awarded tenure there.

So, what caused you to move back to Europe?

I started to become homesick. I am an outdoor person, but Atlantic Canada is quite wild such that it is difficult to get access to the outdoors. You need to have a car for everything, there is hardly any public transportation and there are few signed walking paths. This is unlike Europe.

I began to attend more European conferences and this lead to me receiving job offers. The Max Planck Institute for Meteorology in Hamburg approached me and asked if I was interested in becoming a director there. The Karlsruhe Research Center also approached me to apply for a position leading an experimental research group, with a joint professorship at the University of Heidelberg. I applied for that particular job.

Then came a phone call from a colleague at ETH Zürich asking if I might be interested in applying for the professorship in experimental atmospheric science. This enquiry came again by chance to me, and after the application deadline had passed. So, in a sense, I was being head hunted.

After thinking about it, to start afresh at a new place was very attractive and this position was more attractive than going back to the Max Planck Institute. The position at the University of Heidelberg was based at the Karlsruhe Research Center and was far from the university itself, which meant I would not be near the students, which for me was a negative point.

ETH Zürich as one of the best universities worldwide has the great advantage of attracting excellent students and providing resources to carry out research. At Dalhousie, I discovered that I liked the combination of teaching and research. A director position (as at the Max Planck Institute) instead would have involved a lot of administrative issues and politics. This element of that job was less attractive for me.

I have been at ETH for more than 13 years now and I have never regretted the move. I was asked twice if I would consider becoming a director at the Max Planck Institute, but I was never tempted. I almost felt too young when I got the ETH position and asked myself a few times if there would be a next step? The inner answer for me was "No. This is it."

What have been significant moments that led to this full professorship position?

Throughout my career significant life-changing decisions, or opportunities, have come from chance encounters: a visiting Canadian scholar, the workshop in Canada with the invitation to apply for an assistant professor position, a phone call from ETH, which in fact was a result of my interview for a professorship in Berlin, following my decision not to take that position.

What has influenced your decision making?

I have never moved just for my career and I've always listened to my gut feeling on how I felt about all aspects of my life. I learned to listen to myself and to know when things do not feel right. An early example was that I applied for a postdoc position at the National Center for Atmospheric Research in Boulder Colorado at the same time as I applied for the position in Canada. I chose Victoria in Canada because I wanted to be close to the sea, where I felt more at home.

Another example is that, when I started to feel unhappy at Dalhousie in Halifax, and felt that I needed to move to a university with a better research reputation a colleague suggested that I apply to the University of Toronto. I knew that I could not live in Toronto and be happy, and there was no guarantee that it would be better academically. So, when I got the offer from the University of Berlin, I again listened to myself and realized that it did not feel right to take a position purely based on modeling. I realized that I wanted to have an experimental research focus, even if it meant that I stayed in Halifax for the rest of my life.

Did you have influential mentors?

My PhD supervisor was quite distant and left me mainly alone. I saw him about once every 3 months. What I learned from him, and this was a good lesson, was that if I did not step up and ask for things myself no one would suggest them for me. He made me more independent because of his hands-off approach.

When I got the offer for the postdoc in Canada he was a little bit more encouraging of that idea. He was most helpful when he learned that I was rather unhappy at the government institute in Canada. His advice was that I should not leave one postdoc for another: that is, I should always move up in my career rather than sideways. This advice convinced me to stick it out in Victoria, which meant I was still in Canada, and able to attend that workshop, where I was notified of the tenure-track position at Dalhousie University.

Did you have a dual-career situation to deal within your career?

When I first moved abroad to Vancouver my husband did not move with me. He also did his MSc at the University of Hamburg and then worked for Acer Computers in Hamburg. When I got the offer for the assistant professorship at Dalhousie, my husband quit his job in Germany and joined me in Halifax, Nova Scotia.

As an atmospheric scientist, I had fewer career options, given that he is an IT/software person. However, we took all decisions together and he always vetoed the possibilities or life choices if needed be. He did not really want to leave Canada, but he accepted that I was homesick. I am really fortunate that he came into my life. He is my opposite; he likes to relax, take it easy, and watch TV. He balances my workaholic approach. He originally also studied meteorology, but then moved into IT. When I came to ETH he received a 1-year contract, before he found an IT position at ETH Hönggerberg in the Biology Department. He was treated really nicely in the dual-career move and everything went great.

What has been your experience at ETH Zürich?

It has been wonderful: I have high-quality students and work with world-class scientists. Since the university is directly funded by the Swiss Government, there is independence in our research choices, because not all the money comes from grant applications. Zürich is a great place to be with lots of cultural activities, good public transport and being surrounded by lovely nature. We have a great standard of living at this world-class university. There is also Lake Zürich and I need to be near the water. The whole package is brilliant. Because the quality of life is high, people are happy at work and so productivity is high. The role of a professor is unique and wonderful.

Any highlights of your own research experience?

One really exciting point is that we are putting cloud cameras on gondolas in the Swiss mountains. I am branching into a new field. Secondly, I am able to combine a range of scientific approaches and I lead research both in experimental work and in climate modeling. This duality in research work is only possible in a rich university, because you need a critical mass in a group to make it work. The cross benefits of the two approaches are amazing. The climate modeling gives the big picture, while the lab work provides collected data. I love the combination of working with scientists in the lab and in the field. Modeling is then a tool which can explain what is going on outside.

Were there any obstacles in your career that you had to overcome?

I didn't have large obstacles in my scientific career. Academically I grew up in Canada where there is a flat hierarchical structure, which was really great. Once I got my PhD I never experienced these comments "it's only because you are a woman." In high school the boys could not believe I was better than them and, as an undergraduate, the men could not accept that I was able to explain things that they could not understand. I matured scientifically in Canada and came to ETH Zurich at the top of my career.

Have you had leadership positions at ETH Zürich?

I was head of my institute for 8 years. Until recently, I was a member of the ETH research commission, but I stepped down from it because I was elected to the National Research Council of the Swiss National Science Foundation and will start there in fall 2018. I was also just elected as the Vice President for the Lecturers' Conference at ETH Zürich.

Any other factors in your life?

Rowing is a passion for me. It frees my mind and brings me into contact with the Swiss community and the university community. If you don't have kids you need a way to prevent yourself from becoming a workaholic as a professor. Rowing makes you face the elements. I am outdoors, I do something for myself and it is really challenging in the way it connects so many dimensions—the mind and the body. It has established techniques, and as a college and university sport it requires complex motion, and it always changes. If you pitch yourself against nature it tests you as a person. My husband is a race referee, so this is something that we do together.

I find that the place where I can best clear my mind is in relation to rowing, either during the row or after the workout having coffee or breakfast at the lake.

Prof. Marloes Maathuis (Dutch)

https://stat.ethz.ch/~mmarloes/
Seminar for Statistics, Department of Mathematics, ETH Zürich

Marloes Maathuis, "ETH Zurich/Giulia Marthaler"

Biography

Marloes Maathuis has been full professor of statistics at ETH since 2016. She did her undergraduate and master's studies in applied mathematics at the Delft University of Technology, the Netherlands. She obtained a PhD in statistics from the University of Washington, Seattle, United States (2007). After an additional year at the University of Washington as acting assistant professor of statistics, she came to ETH in 2008 as assistant professor of applied mathematics. She was appointed associate professor of statistics in 2013, and promoted to full professor in 2016.

Research area

Her research interests are in the areas of machine learning, graphical models, causality, and high-dimensional statistics. A particular focus is the study and development of methods to learn cause-effect relationships from complex observational data sets. She also enjoys working on applications of statistics in other fields.

Awards and honors

2017 Elected Fellow of the Institute of Mathematical Statistics
2004 Z.W. Birnbaum Award, Department of Statistics, University of Washington

Conversation with Marloes Maathuis, November 6, 2018

Were you interested in mathematics from an early age?

Yes, I have always liked math, but I enjoyed most subjects at school. Actually, I first wanted to be a medical doctor, like my father. But my high school math and physics teachers questioned this idea, as they thought I would not enjoy courses that involved a lot of learning by heart. They advised me to look at programs focused on thinking and reasoning, like math or natural sciences. My parents, on the other hand, were less keen on such abstract subjects. My father was the first in the family to go to university and valued the idea of having a career or profession, rather than studying a subject with a less defined outcome. They also knew that I enjoyed interacting with people and that did not fit with their image of a scientist. I then considered agrosystems research, which involves mathematical modeling of agronomic systems and land use, so in some sense applied math with very tangible applications—I think that was me trying to reconcile everything. When discussing this with my teachers again, they suggested I just study regular applied math then. This option felt right and the answer had fallen into place.

Another thing I liked about math is that it is used in many different areas. This was great for me, as I did not have to decide on a specific application area yet. In fact, my ultimate choice for statistics fits with these early views. Statistics is used everywhere, allowing me to work on applications in for example biology, social sciences, climate science or medicine. There is a famous quote from John Tukey: "The best thing about being a statistician is that you get to play in everyone's backyard." I fully agree with this and it is an aspect of statistics that I really like.

What was your experience of going to university?

I am from Groningen, in the north of the Netherlands, and chose to study in Delft, which is a few hours away by train. I wanted to experience student life in a new city and I wanted to study at a technical university. I enjoyed my studies a lot; especially the first year was great, with lots of new concepts and ideas and a much higher pace than in high school. I loved learning so many new things and felt like a sponge that absorbed everything in this new environment.

In my class, about 30% were girls. This was quite OK, but we were only 30 people in total. We often had classes together with the computer scientists and the electrical engineers, and the proportion of women in the large lecture halls was very small. I got involved in the rowing community and had a wider group of female friends there, mixing with women who studied industrial design and architecture. We have become a close group of friends, and still see each other a few times a year.

I was also active in the student society for applied math and computer science. I chaired a committee that organized a study trip to the United States and Canada for a group of 25 students, and I took a 1-year break from my studies to take on a full-time position on the board of the society. This is also where I met my partner, who studied computer science.

Your next step was to study for a master's

Actually, everyone in my program went on to do a master's—the bachelor/master split was still very fresh. As part of the master's program, I did an overseas internship. Via connections of my Statistics professor, Piet Groeneboom, I went to Addis Ababa, Ethiopia, where I worked in an HIV/AIDS project. My work involved estimating the lifetime risk of dying from AIDS for people who were born in the year 2000. This lifetime risk is typically much higher than the HIV prevalence (the percentage of people infected at any given time) and it was an important number to communicate the impact of the epidemic. After the 2-month internship, I spent a month traveling, where my boyfriend joined me.

This time in Ethiopia was an incredible contrast to my life before. I learned so much by being in a different culture and loved the whole experience. I wanted more of this. So, when I got back, I investigated the possibility of writing my master's thesis abroad as well.

I thought you did your master's in Delft?

Yes, it was awarded by Delft University of Technology, but I wrote the thesis in the United States. My supervisor in Delft was again my statistics professor Piet Groeneboom—I enjoyed working with him and I found statistics a bit mysterious and difficult to understand, so I was eager to learn more about it. With his help, I could go to the University of Washington in Seattle to work with Jon Wellner. It was easiest to do this by enrolling in the PhD program—it meant I had office space, a visa, and student rights from the beginning, rather than being an unidentified floating person. It also meant I could attend courses with PhD students. I liked this experience a lot, a bit like my first year in Delft. So, after completing my Master's I wanted to stay on and work on a PhD there. The only challenge was that my boyfriend was still back in Delft, and we had to find a solution to the distance. Fortunately, he was able to find a PhD position at the University of Washington as well, working on signal processing for hearing aids.

What was your experience in the United States like?

Seattle is a great place with a lot of cultural diversity and beautiful nature. I lived in a 14-person student housing cooperative for the first few years. We were a diverse group, from different countries and backgrounds, and lived together as a community, sharing groceries, cooking, and maintenance of the house. This was a very fun and formative time for me, with strong friendships and many interesting discussions.

For my PhD, I continued to work with Piet Groeneboom and Jon Wellner. The project involved studying properties of certain nonparametric estimators for interval-censored data. Such data arise in HIV/AIDS studies, so there was still a connection there. After my PhD, I stayed for one more year at the University of Washington as acting assistant professor in statistics. This temporary position was a perfect arrangement for me, as my boyfriend was still finishing his PhD. So in this way I was able to stay in Seattle while continuing my research and getting more teaching experience.

Following this year, you moved to ETH Zurich as assistant professor. How did this swift career progression come about?

My boyfriend and I decided that our priority was to stay together—we did not want a long-distance relationship again. I started to apply for jobs earlier, as he was still finishing his PhD. To maximize our chances of finding a good place for both of us, I applied broadly across the United States and Europe for assistant professor positions. It's not unusual for people in statistics to apply directly for assistant professor rather than postdoctoral positions, because the job market is tight and universities want to recruit people before they go somewhere else. I had options in the United States, in the Netherlands, and at ETH, and we decided to go to Zurich.

What made you choose ETH?

ETH has a great statistics group, very good students, and stable funding. Such conditions are hard to find elsewhere. Also, ETH included a 1-year postdoc position for my partner, and the dual-career office helped to establish contacts with groups working in his area. We really appreciated this and felt welcomed. Finally, Switzerland seemed a very nice country to live in, and we were happy to be closer to our families in the Netherlands again.

The only disadvantage of the ETH offer was that it was non tenure-track. This is a policy of the Math Department, to employ assistant professors on short-term contracts only, which meant my contract was for 4 years initially, with a possible 2-year extension. We were not thinking very far ahead at this point, so we did not reflect on this too much. We just felt that Zurich was the best place for us for the next few years. Also, I was never fully set on an academic career, and therefore the temporary contract didn't feel like a very large risk.

You started your family in the early years after you came to ETH

Yes, we arrived in 2007 and our sons were born in 2009 and 2011. We did not want to wait too long to have our family. My work environment also played an important role here. There is a healthy work-life balance at my institute. My colleagues also advised me to live my life and not put everything on hold until I would get a tenured position. This support, together with the fact that I could see interesting careers outside of academia, and some basic trust that things would somehow work out, allowed us to make this decision. I am very happy that we did.

How would you describe your experience of having a young family at ETH?

A great aspect of my job is its flexibility. I can work from almost any place and at any time. For example, I thought about research problems while walking around with our babies in a sling. And now that the children are at school, I can be with them during an afternoon or attend some event at school, and then make up for this some other time.

When our sons were younger, they were both in ETH daycare, in the KIKRI ETH Zentrum. This is a wonderful place—it is a warm community for both parents and children and was a great place for our children for their first out of home experiences.

Your partner started at ETH on a 1-year postdoc contract, given as dual-career support. What happened after that?

Initially, he worked at the University Hospital on a project related to signal processing for cochlear implants. Afterwards he applied successfully for postdoc funding related to industrial applications, so he could continue this research position there. He had a lot of freedom, but it was a very small group. So, in 2011 he decided to go into industry. He applied successfully to join Google at their European headquarters in Zurich, where he is very happy. He enjoys this work environment more than academia.

Your assistant professor position was for a maximum of 6 years, with no tenure-track route. How did the transition to full professor at ETH come about?

Toward the end of this period, a new professorship in statistics was created. It was an open call, and I found it a bit awkward to apply as internal candidate, but decided to go for it anyway. I wasn't offered this particular position, but given that there was a professor in our institute near retirement, the ETH President at that time, Ralph Eichler, decided to appoint me as an early replacement for this position. This is how I became associate professor in 2013. The step to full professor in 2016 went through the regular promotion process.

The step from the non tenure to the tenured position was not an easy process, but in the end, it worked out well for me. I was very lucky with the timing, in the sense that a new position was created and, at the same time, a professor was close to retirement.

What is the difference in your position now?

As far as research goes there is not a lot of change. I continue to get grants from the SNSF for 4–5 years funding. The main difference is probably increased responsibilities. For example, I am now study director of the master in statistics program, I am on the steering committee of the master in data science program, and I am more involved in the organization of conferences and workshops. I also sit on many more committees and am taking on more responsibilities in professional societies.

Could you tell me a little about your research interests and projects?

After coming to ETH, I shifted my research to the fields of causality, graphical modeling and high-dimensional statistics. I think understanding causal relationships is at the basis of science. It is relatively easy to do this from so-called interventional data, where researchers conduct an intervention. For example, if you see a change in the growth of plants after adding a certain nutrient, then it is easy to infer that the nutrient affects plant growth. Observational data are data that are just recorded without actively intervening. From such data, it is easy to infer whether there is a correlation between certain variables, but correlation does not imply causation.

For example, I am currently working on a project to investigate the causal effect of caffeine intake on sleep quality. Even if one saw a correlation between caffeine intake and sleep quality, say that people who ingest more caffeine tend to have a worse sleeping quality, then this would not imply that caffeine negatively affects sleep quality. The main problem is that the groups of people who ingest a little or a lot of caffeine differ in many respects. For example, the high-caffeine group may tend to have higher stress levels, and this may affect their sleep quality. Together with lots of other variables, including genetic data, we investigate what we can and cannot say about the causal link between caffeine and sleep quality under different sets of assumptions.

The project about caffeine and sleep quality is on the applied side for me. I also have more methodological projects, where we develop new methods, study and derive properties of new or existing methods, or try to improve computational bottlenecks.

What would you say made the difference in your career?

It was certainly people—my family, my partner, my friends, all my teachers and mentors, and also my students and collaborators.

And of course, my statistics colleagues at ETH. The atmosphere at our institute is very good, and this makes it fun to go there everyday. It was also very important for me to see that my senior colleagues were highly successful in their work while having a good work-life balance. It really mattered to me that I could be a professor and also have a personal life and a family. I think it was key that I saw that this is possible, whether for males or for females.

Also, I realize that coincidences have played a big role. Each little step in my career affected the next, and if one of these steps would have been different, it might have all come out differently.

What other factors are important?

My partner and I are fully sharing the responsibilities for our boys and household. He is super-supportive and that makes it all possible. During the school holidays, we sometimes have help from our parents in the Netherlands, who come over to spend time with the children for a week.

How would you summarize your job?

It is really interesting, every day is different, and it involves a huge range of tasks. I enjoy the mix of research, teaching and organizational aspects, and I love working with people. So what my parents predicted came true—I am a people person—and I am glad that I can use this everyday. I also feel privileged for the freedom I have. It is great to be able to work on the problems that fascinate me.

Prof. Isabelle Mansuy (French)

http://www.hifo.uzh.ch/en/research/mansuy.html
Brain Research Institute, Medical Faculty of the University Zürich (UZH), and the Department of Health Science and Technology, ETH Zürich

Isabelle Mansuy

Biography

Isabelle Mansuy is full professor in neuroepigenetics (since 2013) at ETH Zürich and University of Zürich. She completed a PhD in developmental neurobiology at the Friedrich Miescher Institute in Basel, Switzerland and the Université Louis Pasteur Strasbourg, France. From 1994 to 1998 she worked as postdoc in the lab of Eric Kandel, Center for Learning and Memory, Columbia University, New York, United States. She was appointed assistant professor in neurobiology at the ETHZ in December 1998. In 2005 she became associate professor in molecular cognition at both University of Zürich and ETH Zürich.

Research area

Isabelle Mansuy's research examines the epigenetic basis of complex brain functions in mammals and focuses in particular on the mechanisms of epigenetic inheritance. The goal is to determine the processes underlying the influence of life experiences on mental and physical health across generations. This work recently demonstrated that early trauma in mice induces behavioral and metabolic alterations and that these alterations are transmitted across several generations through the germline. Her research is conceptually and technically original and innovative, and the lab is a pioneer in the field of transgenerational epigenetic inheritance, a discipline at the forefront of a paradigm shift in genetics. The lab is also conducting collaborative studies in trauma patients with several psychiatric clinics in Europe to assess the relevance of its findings in mice for humans.

Award and honors

Isabelle Mansuy is member of the Swiss Academy of Medical Science, the European Academy of Sciences (EURASC), the Research Council of the Swiss National Foundation, of the Research Council of the Fyssen Foundation and of EMBO, and was elected Chevalier dans l'Ordre National du Mérite (2011) and Knight of the Legion of Honour (2016). In 2017 she was elected as Fellow of the European Academy of Sciences.

Conversation with Isabelle Mansuy, July 20, 2015

My research field is exciting and not classical, and it gets lots of public interest. It was a challenge to establish myself as professor in this new field at first, but it has now become a wonderful although tough job with independence and freedom.

Where did your interest in science come from?

I always liked physiology, though I come from a modest family from the county of Les Vosges in France. My father was a working man, my mother stayed at home. I am one of five children, we lived on a low income, but were given strong moral rules: work hard, be honest, respectful and become independent. You need to drive your own life through your own work.

We began to work early as young children. It was our duty with tasks in the house and on the farm of our grandparents. It was not an unhappy childhood, but it was also not particularly playful and there was little leisure. All of us had bright minds. My father is a self-made man who can do anything, build a house, create paintings or bake a cake, and is driven by passion. He is still very active even if now 78 years old.

As a child, I was bothered by authority and my father was very authoritarian. I had great capacity to work without complaining and I was working for my freedom and independence. This capacity stood me in good stead through the tough times, because I can tolerate a lot. I now have a privileged job, a wonderful life, as a result of all the hard work.

How would you outline your research field?

I focus on the mechanisms which explain how life experiences can affect mental and physical health in exposed people and their descendants. We focus in particular on traumatic experiences because they are known to be associated with severe psychiatric disorders and because such disorders are often passed down in some way to children and grandchildren even if they never experience any trauma. A good example is the long-term effects of war on mental and physical health. This topic really speaks to the general public.

Clinicians already know that trauma and mental health problems run in families and that this is not only due to genetics. There have been many longitudinal studies by psychiatrists in the past decades, but what explains such transmission is not understood. The most prominent hypothesis is that transmission involves epigenetic factors present in gametes, which are the oocytes and spermatozoa, that are altered by exposure to trauma. These alterations are thought to be transmitted to the descendants, just like genes are, and explain why similar trauma symptoms can be manifested in children and grandchildren.

Such transmission of the effects of life experiences is a general concept that is very important for human beings. It is called epigenetic inheritance, and applies not just to trauma but also to nutritional insult, exposure to endocrine disruptors, smoking, etc. Among the major studies in humans is the assessment of the impact of famine or very low caloric intake in people and their descendants in Sweden and the Netherlands. The results showed an increase in the incidence of metabolic diseases, obesity, diabetes and an increased mortality in these people.

I am interested in the concept of epigenetic inheritance related to brain functions and its importance for psychiatric diseases like depression, borderline personality disorders and their heritability. These conditions are known to be associated with traumatic experiences in early life and result from a combination of genetic and epigenetic factors. To complement the experimental work that we are conducting on animals in my lab, I am collaborating with psychiatrists in Switzerland and abroad. My research with mice has demonstrated that exposure to trauma in early postnatal life, such as the chronic and unpredictable separation of pups from their mother combined with maternal stress, has an impact on behavior and metabolism across generations, and that this is likely caused by epigenetic alterations in sperm cells. Working with a mouse model is extremely important because it is not possible to do such studies in humans. However, the epigenetic markers that we identified as being associated with trauma in mice can be examined in human blood to determine if they are also altered. They are expected to be very useful biomarkers of trauma.

So how did you go from the countryside in France to becoming a professor in Switzerland?

My parents wanted us to be professionally active as soon as possible but I wanted to leave home and take on the highest intellectual challenges possible. I took preparatory classes for 3 years to get into an engineering school, or Grande Ecole, and afterwards entered a new university in Strasbourg, which was offering education in genetic engineering, a novel discipline at the time. The school was a combination of a university and engineering curriculum. It was an international school, very intense and exciting with passionate teachers and open-minded researchers. The teachers came from all over Europe, this gave me the drive to work abroad after my degree. I left home at 14 and 1/2 years old, my older sister left at 15. It was challenging to do this, but we succeeded. My parents supported my choices, but I had to make them happen financially and I was very determined.

For my PhD I was interested in new technologies and wanted to work on transgenic mice. I did an internship at Novartis in Basel for 2 months, and then did a PhD in developmental neurobiology at the Friedrich Miescher Institute in Basel together with the Université Louis Pasteur in Strasbourg. I realized that neuroscience was the research area that interested me most but I also wanted to work in psychiatry.

Your next step was a postdoc position in the United States?

I applied to several labs in the United States though I was not sure if I would get the position I wanted. When I visited the lab of Eric Kandel at Columbia University, something clicked and I knew that it was the right place for me. Prof. Kandel agreed that I join the lab but I had to choose and conduct my own project and if I could, support it financially. I had the independence and freedom I wanted but it was really tough. When I started at Columbia I had a 4-week-old baby. When I'd met Eric Kandel, I was 6 months pregnant and this was no problem for him, which made me realize that he would probably be a good boss.

How did you manage this change of continents, groups and having a small baby?

My husband worked for the French government as a mechanic, so he was able to take a sabbatical to move to the United States. It was not easy as sabbaticals are not usually allowed for people with his job, but we found a way. He stayed at home and looked after our daughter. It was a brilliant combination, though it was not easy for him to do this.

I stayed at Columbia for 4 years and then was offered a research position in Paris at the end of my postdoc. However, at the same time, I was also offered a non tenure-track assistant professorship at ETH Zürich, which I accepted. My husband had however accepted a wonderful job in Paris 6 months earlier, and he had to work there for a year to meet the criteria for his contract with the French government.

This meant that in December 1998 I moved to Zürich with my daughter and the family was separated. We lived in a one-bedroom apartment and I had no access to childcare at ETH. She went to the Lycée Francais in Gockhausen, Zürich for 6 months, but it was a very difficult situation. My daughter was unhappy because she was not with her father, who she was very close to and, after 6 months, we decided to move her back to France. But because the separation was too harsh for us, my husband decided to take another sabbatical from his job, which was absolutely vital as we could not have continued with such long separation. We then decided to establish the family home in France near the Swiss border, and after 4 years, my husband started to work again. This meant that the family was still not together as I was in Zürich during the week (2 hours away), and at home in the weekend. We have been living that way since then.

My husband was a wonderful dad. My daughter is now 24 years old and is finishing a masters in optometry in Paris. Her father was solid and has always been extremely supportive through the time period when I set up my lab and thereafter.

You had a non tenure-track assistant professorship, which meant no career track at ETH. How did find a way forward?

I had a 6-year contract and after 2.6 years published two big papers, which was very important. My lab was quite successful, though I was always under pressure. After many discussions, I was offered a joint professorship between the University of Zürich and ETH, which was a good combination but the conditions were not optimal and the financial support was minimal, and even not existing from the ETH for 2 years. I was made associate professor for molecular cognition (UZH/ETH) in 2005, then full professor in neuroepigenetics in 2013, but despite reaching this position, the situation has always been very difficult because of a suboptimal financial support.

Did you always want to be a professor?

Absolutely not as I did not even know what it was to be a professor. I never really had specific career goals, but wanted to be professionally free, and do something exciting. I also wanted to do challenging research that was useful for medicine. I would not have survived the tough times without the exciting research, and there is so much more to do in this field.

I am on the forefront of research where techniques are constantly being challenged. It is great to work with psychiatrists and to have contact with the public. I am often invited to give public talks, which is always very interesting and enriching. For example, I was invited to the "City of Vienna" to talk about inherited violence in young people. I teach Epigenetics and Neurogenetics in a range of universities and to the public.

The greatest thing about a professorship I think is intellectual freedom. I do feel the pressure to produce good research, to work in a competitive environment, and make it visible, but at the same time, it is extremely rewarding to know that medicine and people can benefit from it. For me, this job is a privilege and I feel quite invested by it.

Prof. Paola Picotti (Italian)

http://www.imsb.ethz.ch/research/picotti.html
Institute of Molecular Systems Biology, Department of Biology, ETH Zürich

Paola Picotti, "ETH Zurich/Giulia Marthaler"

Biography

Paola Picotti was appointed associate professor for molecular systems biology at ETH Zürich in October 2017. She was born in Udine, Italy, in 1977. She studied pharmaceutical chemistry and technology at the University of Padua in Italy and completed a PhD on the molecular mechanisms of protein folding at the same university's CRIBI Biotechnology Center in 2006. She then joined ETH Zurich as a Marie Curie postdoctoral fellow in Professor Ruedi Aebersold's group, Institute of Molecular Systems Biology, where she played a major role in developing targeted proteomics technologies. From 2011 to 2017 she was assistant professor (SNSF Professorship) of biology of protein networks at the Institute of Biochemistry (IBC) in the Department of Biology.

Research area

Paola Picotti is interested in how protein conformational changes impact cellular networks and result in physiological modulation or disease. Her group studies the molecular bases of protein aggregation diseases, primarily Parkinson's disease. A strong focus of the research is also the development of mass spectrometry-based structural and chemical proteomic methods. A recent contribution of the Picotti group was the development of an innovative structural proteomics technology to probe protein structural alterations directly from cells and tissues. Paola Picotti is also involved in the development of publicly accessible tools to promote the dissemination of proteomics approaches and chairs the working group on protein aggregation diseases of the world Human Proteome Organization.

Awards and honors

2018 Juan Pablo Albar Protein Pioneer Award
2018 Friedrich Miescher Award
2016 Robert J. Cotter Award, American Human Proteome Organization
2016 SGMS Award for independent research on mass spectrometry
2016 EMBO Young Investigator Award
2014 ERC Starting Grant
2014 Top-40-under-40 Power List, *The Analytical Scientist*
2011 SNSF Professorship and Latsis Prize

Conversations with Paola Picotti, June 11, 2015 and 2016

When did you first become interested in science?

I was always interested in science and technology, particularly chemistry, biology, and math. Math was actually my first love at school, and I even won regional competitions at high school. I chose to study Pharmaceutical Chemistry at the University of Padova. In the first 3 years, we studied basic chemistry and biology, but after that the objective of the course was to design new drugs.

I remember a moment in one of the big lecture theatres when our biochemistry professor showed us a slide with the structure of a protein. It was the ATP synthase, the enzyme that generates the energy-storage molecule ATP (Adenosine triphosphate). I was truly fascinated by its complex structure and at that moment I felt that I would love to study proteins my whole life. Studying proteins is the topic of my current SNSF professorship—I am studying the structure of proteins and how altered protein structures result in disease. Seeing the ATP synthase for the first time was stunningly beautiful—in textbooks there are lots of illustrations of other proteins such as hemoglobin, which also show you how complex they are.

What was the result of this personal insight?

I did a master's thesis on proteins, and I started my PhD at the Center for Interdepartmental Research in Innovative Biotechnology, University of Padova. During my PhD, I took the initiative to go to ETH Zürich as a visiting student to work on proteomics and on mass spectrometry. This experience made me realize that at that time Italy was not strong enough technically for me to do the research that I was interested in.

I'd organized a research period in Prof. Ruedi Aebersold's lab and, though I had read many of his papers, I had not realized that he had such a worldwide expertise and reputation until I was at ETH. The decision to become a visiting student there was the best I ever made and I worked in the Aebersold lab for more than 9 months, during the last stages of my PhD. I had wanted to be exposed to a new environment, and it was life-changing, because I moved my research focus from the analysis of single proteins to the technological development of tools that enable studying simultaneously the ensemble of proteins in our cells or proteome.

I returned to Italy to defend the thesis and then was offered postdoc positions in three labs: in Padova at CRIBI, a position at the University of Udine, and at ETH

Zürich. Because I was in a relationship in Italy, I took the Udine job for a few months, it was real decision-making period for me. Ultimately the postdoc was not rewarding, they did not have the technical facilities nor the expertise for the research that I wanted to do. At the end of 2006 I took up the offer to join the Aebersold lab as a postdoc and then spent 4.6 years in Prof. Aebersold's group.

How would you summarize this time at ETH Zürich?

It was a life-changing period in two ways: Prof. Ruedi Aebersold was an incredible mentor and, secondly, I met my husband in the same lab.

Aebersold changed my view of science. His approach of focusing on transformative science rather than on incremental improvements was truly inspiring. He encouraged me to be brave and pursue new concepts, which was wonderful, because this fitted with my character. It was illuminating to do the research with him because the research we did changed the way that protein analysis was done at that time. The field was working on optimizing an existing method, but the Aebersold lab took a new and completely different perspective. It was hard at first, we felt the opposition in the community. Currently, the method we developed at that time is considered "the gold standard for the targeted mass spectrometric analysis of proteins." It is called selected reaction monitoring and in 2013, Nature Methods selected it as "method of the year."

This work is the reason I won the Latzis Prize in 2013. This method can quantify specific proteins in a complex mixture. We are able to target one protein out of a complex mixture of proteins by using a mass spectrometer.

What were the elements of mentoring that made a significant difference?

Prof. Aebersold had a really excellent and subtle way to support my career. He cultivated me and helped me grow, probably because he spotted my potential. He threw me into deep water, and gave me responsibility for the most challenging (but at the same time very rewarding) projects in his lab. He also linked me to the people in his network in the United States and Asia. He challenged me in a way that helped me grow. He encouraged me to lead large projects and built my self-confidence by giving me challenging tasks and the tools to take the next steps.

Ruedi did not really praise me directly, but he nominated me for many prizes and gave insightful advice every time I requested it. He kept a distance and did not have personal conversations with people about my applications for the next stage. He wrote all the references to support my career steps, but he let me do the rest on my own.

Your next step was to apply for an SNSF professorship. Why did you remain at ETH?

Yes, this was a risky choice, because I had to establish myself independently from my mentor. When I made the application for the SNSF professorship I could have taken up the position at the University of Zürich, the University of Geneva, as well as at ETH Zürich. In the latter case my support was from the Institute of Biochemistry, rather than the Institute of Molecular Systems Biology where I had been based. None of these universities had open calls for professorships, so to pursue my career, the SNSF route was the best way forward.

Once I received the award I decided to stay at ETH because of the scientific fit, the ETH facilities and my focus on technology. I was very aware that it was necessary to separate from my mentor, and I had changed Institutes. I knew that comments would be made in the department if I did not do this. I developed a completely independent project and avoided overlaps. It was very difficult initially, but also extremely rewarding after 4 years. The key for me was to make a complete paradigm shift, and not to develop my research incrementally.

I asked myself which characteristic of proteins could be used as an informative readout for pathological states, more informative than their abundance, which I had measured as a postdoc and which everyone else measured in the field at that time. The main outcome has been novel technology, based on mass spectrometry, for examining altered structures of proteins and directly from human specimens. Protein structures integrate various cues that affect protein function and are thus very powerful biological readouts. This technology opened up a number of new avenues. There are many human diseases where protein structures change and there needs to be biomarker discoveries—for example in Parkinson's and Alzheimer's diseases in which changes in proteins can lead to neurodegeneration. Our technique can be used to track protein changes in blood and tissues and potentially spot the changes early. It also supports basic biological research because detecting altered protein structures can shed light on the function of a protein.

You recently had your first son and, at the same time, you and your husband are managing a dual-career situation. Were there any challenges during this time?

During my maternity leave I kept working on papers from the lab and published papers in Nature Biotechnology and Nature Methods. My colleagues did not expect productivity during my maternity period and were surprised about this. One really wonderful outcome was that, during my maternity leave, my department proposed me to the university's governing body for a bonus because of this productivity. The official recognition was really worth it, and my lab received a bonus of 100,000 CHF from Prof. Ralph Eichler, the president of ETH at that time. I hope that, as a result, I am changing stereotypical attitudes.

Just before my maternity leave I was delighted to find I'd been awarded an ERC starting grant. In fact, my husband, who is an assistant professor at the University of Zürich was also awarded one at the same time and we had competed on the same panel. We planned to try and make this to happen alongside my pregnancy and everything went remarkably well.

During my maternity leave it was important to keep in constant touch with my group. I regularly spoke with them over Skype and joined group meetings via Skype as well. I was virtually present, but did not go into the lab physically. Another factor that helped with the maternity leave was that I had hired a highly specialized scientific lab manager before I went on leave. I used the budget I received from my host institute, which would normally go toward hiring two PhD students. Until that point I was the technical expert for all our big instrumentation in the lab (the mass spectrometers), now my lab manager can maintain and operate the three mass spectrometers, and he did this throughout my maternity leave. He also kept some of the external research collaborations going, while I supervised the science. He continues to work for me and it has been a very effective solution.

What about the return to work? How did you manage that?

I finally got a spot in the ETH daycare just before I was due back at work. I think the facilities need to be expanded. They have now established a baby only group, which is great, but there are real problems, with very long waiting lists and it is not clear how they allocate places. (*In 2016, a second daycare facility was opened on the Hönggerberg Campus*). My son is now 1 year and 4 months. We had an initial chaotic phase, especially when dealing with emergencies, such as sicknesses of our son, but now things are under control. My husband is very supportive and we do everything 50:50 down to the smallest detail. In addition, our grandparents live reasonably close (Lake Constance and Udine, Italy) and can come to help. At this point I have begun to increase my presence at conferences again and I am traveling more with the help of my parents.

A further factor that helped was saying "no" to certain responsibilities in the first months of returning to work. I needed to focus on my lab and ensure that the research continued. I simply had to learn to say "no." A supportive colleague, who had already gone through the process of becoming a parent, advised me not to take on extra commitments during this time. I rejected an offer to teach immediately on my return and I turned down the possibility of sitting on some of the Department's PhD committees.

You said that your husband is an assistant professor at the University of Zürich, so there is still the challenge of you both finding permanent positions

My husband was a PhD student at ETH when I joined Ruedi's lab as a postdoc. When he finished, he went to Stanford University to do a postdoc and developed a new technology in the field of protein analytics. He then was hired as assistant professor, non tenure-track, at the University of Zürich. There will be a new professorship there in 2017 in Quantitative Biology, which he is able to apply for, but given that our professorships at both universities are not tenure-track there is no possible discussion of a dual-career situation.

My contract would normally be for 7 years, but the successful ERC grant means that I have 8 years. The next step for me and my family is for both of us to get permanent professorships. We are working in similar fields, which narrows the options. We will, I think, need a "dose of good luck." A full professorship is coming available in my department when my former mentor retires and I may apply for that position.

What would you say were the challenges in your career so far?

My first challenge was my transition to the Institute of Biochemistry where the majority of people do Biology related research, while mine was more technology based. I needed to learn a different language, and I had to read a lot and attended different conferences outside my field. Initially that community did not necessarily acknowledge the importance of the development of new technologies, however, after 5 years I am well integrated, I have many collaborations and they see the enormous utility of our approaches. It has been an enriching experience and I would suggest everyone to take, at least for some time, a detour into a different field.

Another challenge was to attract good postdocs. As a young professor, I was able to attract good PhD students and master's students, but it was harder to get good postdocs because I was an unknown researcher. After a successful ERC starting grant and a few publications on high-ranking journals, great people started to apply. Initially, I received spontaneous applications from male postdocs only, this is perhaps because of the technology focus of my research (although it should not be like that!). After posting a Nature advert, I got 80 applications and invited 10 for interview. There were two women out of those 10 applicants and both were outstanding. I hired both women to increase the diversity of the lab.

I never experienced discrimination, but there were the usual inappropriate comments. Every time I won a grant or a prize there were always people who said "of course you got it because you're a woman." This is really a frequent experience. It made me check how the ERC grants are actually awarded and I found that there is absolutely no element in the ERC grant scoring system that includes explicit favoritism toward women, so those comments are completely unfounded.

I was once told by a departmental colleague that my appearance is too soft and that I will have to change my character to be successful. I haven't changed my character, but I have developed management instruments and techniques. I can remain soft and friendly, but tough when needed, it is all about the techniques. I have learned that it is very important not to become too emotional. I have developed a professional role, used techniques, but fundamentally remained "a nice person." I plan also to do the well-known lab management course of the European Molecular Biology Laboratory (EMBL), which has useful mentoring associated with that course.

It's 2016 and I hear there is news on a professorship at ETH Zürich!

Yes, I have gone through a challenging, but very positive process in my department. The aim of the process was to find on a worldwide scale the best person for the research profile that had been defined by the faculty of the department. The profile of the new professor was in the field of proteomics and meant as a replacement of my mentor Ruedi Aebersold who will soon retire. After a thorough interview and extensive external reviews of my research, my department decided that I should be nominated as the successor of Ruedi, thus making an exception to their non tenure-track system. Due to this exception and because in-house recruitments are generally rare at ETH, the selection process was particularly rigorous. One hundred percent of the professors in my department needed to vote for the proposal, and there had to be a unanimous vote to accept me as this candidate. The vote was unanimous in my favor, which was absolutely brilliant. Initially it wasn't clear whether I would be appointed as an associate professor (i.e. with tenure) or have to go through the tenure-track process as well. The ETH president, Lino Guzzella, decided that the assessment process I'd been through already to reach this unanimous vote by members of my field, had been very rigorous and it made no sense to go through a second similar assessment for tenure. If all goes smoothly the negotiations take place in Fall 2016, I will be appointed in 2017, and move back to the Institute of Molecular Systems Biology, where I was a postdoc.

In the meantime, my husband was appointed as associate professor of quantum biology at the University of Zürich, and I am currently expecting my second child by the end of the year. This is all truly tremendous.

In October 2017 Paola Picotti was appointed associate professor for molecular systems biology at ETH Zürich. In early 2017, during her maternity leave, and after the arrival of her second son, Paola's research on proteome structure was published in Science and later on in Cell. Her research in this field was awarded the "Protein Pioneer Award" of the European Proteomics Association, the Friedrich Miescher Award of Life Sciences Switzerland, the EMBO Young Investigator Award and the Rob Cotter Award of the American Human Proteome Organization.

Prof. Ursula Röthlisberger (Swiss)

https://lcbc.epfl.ch/roethlisberger
Laboratory of Computational Chemistry and Biochemistry, Institute of Chemical Sciences and Engineering, EPFL

Ursula Röthlisberger, “Alain Herzog/EPFL”

Biography

Ursula Röthlisberger has been a full professor of computational chemistry and biochemistry since 2009. In 1988 she completed her diploma in chemistry at the University of Bern, then from 1988 to 1991 she studied for a PhD working at the University of Bern and IBM Zürich Research Laboratory. She continued for a year's postdoc at IBM/Uni Bern and then moved to Michael Klein's group, the University of Pennsylvania as postdoc fellow from 1992 to 1995. She spent 1995–96 as a postdoc with Prof. Michele Parrinello at MPI Solid State Physics, Stuttgart, German. After 1 year as Profile 2, SNSF Fellow, ETH Zürich she was appointed as assistant professor of computational inorganic chemistry from 1997 to 2002. She moved to EPFL and became associate professor from 2002 to 2008.

Research area

Ursula Röthlisberger's research interests are concentrated on ab initio MD methods based on density functional theory (Car-Parrinello simulations) and their application, adaption and extension to systems of chemical and/or biological interest.

Honors and awards

2016	Doron Prize
2015	EuChemMS Lecture Award
2015	Elected as member of International Academy of Quantum Molecular Science
2004–07	Member of Research Council of Swiss National Science Foundation
2004	Dirac medal for “outstanding computational chemist in the world under the age of 40" from the World Association of Theoretically Orientated Chemists

Conversation with Ursula Röthlisberger, October 19, 2015

When did your interest in natural sciences begin?

As a young person, I found it difficult to choose my favorite topic because I loved doing everything, but I liked philosophy and the exactness of science particularly. I was drawn to the fact that these subjects enable you to understand the world. In addition, my sister had studied chemistry already, and I saw how the topic brought physics and biology together. I wondered how it would be to study them and I became interested in exact sciences and thinking about how to apply physics and mathematics to biology.

Certain other factors shaped my schooling decisions. I come from a small village and I could have gone to the gymnasium at 11 years old, but my mother wanted me to stay longer in the village school. By the time I moved to the gymnasium it was no longer possible to study the older languages such as Latin and Greek, because you had to choose them earlier, so I focused on the science specialism for the Matura exams. I could have chosen Art or Literature but I was very idealistic and wanted to understand the origins and ideas behind how the world works. I was always asking big questions and began to find that philosophy gave only vague answers. Scientific topics gave meaning and provided answers.

How many girls studied the science Matura?

There were definitely less girls in my science classes, but it was when I went to the University of Bern I really noticed the difference. When I began there were only six girls out of 100 students and, after the first semester, the rest of the girls gave up their studies. I continued and followed my degree with a master's course in experimental research using lasers and I was the first young woman to work in the research lab. At the same time, I was also introduced to computational methods.

This must have a been a significant step, given that you became a theoretical chemist?

At this point I was working in an experimental group, but we used computational methods. When I studied for my PhD in Bern I also had the opportunity to work at IBM's research labs in Rüschlikon, for 4 days a week, with Wanda Andreoni and Michele Parrinello who were there working on fundamental research. Prof. Parrinello had just joined IBM and it was an amazing time to work there, and a particularly high time for IBM in terms of research. We lived in the lab, we socialized together as a group and then returned to the lab. One day a week I did experimental work at the University of Bern and I was also a teaching assistant there.

You completed your PhD in record time. How did that happen?

I did my masters in the lab of Prof. Ernst Schumacher, who was a very charismatic supervisor. I wanted to continue with my PhD in his group, but I had to agree to complete it in the years left before his retirement. I worked day and night, and it was done in 2.6 years.

What were your next steps?

My boyfriend was still finishing his PhD in another group so I then worked for 6-month as a postdoc at IBM. Afterwards we both applied to universities in the United States to see if we could find positions where we could both do research work. In the end, there were three to four places where we had opportunities. I was offered a position for 2.6 years and my boyfriend was awarded an SNSF fellowship to fund his research. At this point I decided to focus on pure theory, even though I was told that there was no way that I would find an academic position in Switzerland afterwards as a theoretician. I was dampened by this attitude, but I went ahead and did it anyway.

Where did you do your postdoc in the United States?

We were based in Philadelphia, which was not a safe city and was a real change of lifestyle from Switzerland. My supervisor, Prof. Michael L Klein had ongoing collaborations with Switzerland, which enabled me to maintain my research links back home. Michael Klein was very supportive and even encouraged me to apply for an academic position, or assistant professorship, very early. This made me start to ask the question, "What do I want to do for my future research?" I was invited to interview for an open position at the University of Rochester and, though this professorship was in the end canceled, and I did not interview, the whole experience made me clearer and determined about the direction I wanted to take. The fact that I had made it to the shortlist on a first application was deeply reinforcing.

How did you manage your move back to Switzerland?

I applied successfully for a Profile II fellowship from the SNSF, to be based at ETH Zürich. I was still very young at the time, and one condition of my acceptance, was that I would able do one more year as a postdoc before starting the fellowship. I wanted to learn more techniques and thus I joined Michele Parrinello's group at the Max Planck Institute in Stuttgart and stayed there for a year. Afterwards I moved to the group of Prof. Wilfred von Gunsteren at ETH Zürich. At the time, the status of Profile II fellows was very ill defined, they could be hired by the host institution as senior scientist (Oberassistant) or just as simple postdocs. After 6 months, I realized that the local computers were not good enough for my computational work and I wanted to apply for grant money to buy additional machines. My host professor said "No" to supporting me with my application, which was a disappointment.

I discussed my options with Michael Klein in Philadelphia and he advised me to apply for positions in the United States. I applied for an assistant professor (AP) position at Princeton and got the job. ETH Zürich then responded by informing me that there was also an AP position open there, and my application for the AP in Zürich was successful too. After some negotiating back and forth I decided to stay at ETH for family reasons, moving from physical chemistry to inorganic chemistry, with my research focused fully on computation. This was really good outcome from a period of questioning and reflection.

It made me realize that sometimes a challenging experience can, actually, be a very good thing. I was told not to apply for a grant that I needed and, as a result, it made me open up my options and think more broadly. It gave me the incentive to think about moving on. As long as you are not caught in a position and are able, and willing, to leave, then you are in charge of your choices. However, if you are tied to a particular city or country it can be very tough. Going to the United States was always an option for me, a real option. At this point I no longer had a dual-career situation because my boyfriend and I had split up in the United States.

What was your experience of being an assistant professor?

There was no option for tenure in my position at ETH, unlike if I went to the United States. After 2 years as an AP, I received an offer for a full professorship position at TU Berlin. It was made clear to me that there would be no position for me in Inorganic Chemistry at ETH Zürich, though there was a professorship opening up in Physical Chemistry, but it seemed that they wanted a person in a specific field, which was not mine. My colleagues congratulated me on my move to Berlin, even though I had not accepted the position.

What did you do?

I turned down the professorship at TU Berlin because I had only been at ETH for 2 years and was just starting to put my group together. Another factor in my decision was the fact that TU Berlin did not have enough resources to replace the computers I needed for my research. My colleagues could not understand my decision, but afterwards, I received offers for full professorships in the United States, from the University of Wisconsin-Madison and for the Quantum Theory Project (QTP), University of Florida.

Next, I was offered a tenured professorship at EPFL, Lausanne, in the French side of Switzerland. This professorship came with full equipment, funds, and wide possibilities—and I could move my group easily there. It was a very good offer and really the best option for me. I became associate professor at 36 years old, which was relatively young. I appreciated that EPFL put their trust in me and believed in me. Originally, they'd created a junior professor position, but they turned it into a tenured position when they appointed me. It was the first theoretical chemistry position at EPFL, and now they have six theoretical positions. I got all I asked for and more. There was a great atmosphere there (like in the United States), with excellent young people, and good generous support. It would be hard to find a better place than EPFL.

How did you find your central research strategy or direction?

I was always convinced about what I wanted to do. Parinello's computational methods were not used in chemistry and biology and I particularly wanted to apply them to biology. I was not convinced that I could make it and be a success, but I always knew that this is what I wanted to do. I'd had some experience teaching at a gymnasium and I'd worked three times in industry during semester breaks, so I had experienced a lot of other career options. I also realized that doing my research connected my heart and mind together, and this awareness helped me to deal with obstacles in my career, or difficult politics. I was not prepared to give up my academic career easily because my choice felt right. I saw difficulties or problems as an incentive to do something to counteract them and to find another way to solve the issue.

How would you describe your experience of being a full professor?

I have been doing this for 16 years now. Once I moved to EPFL I could focus my research on Biology, which was really great. However, in an academic career, you need many more qualities in order to lead a group than those you are introduced to at the level of PhD and postdoc. You also experience more acceptance into the community as a junior level professor, but the challenges grow as you become a more senior professor. You need to be prepared for this and find strategies to deal with it.

What would you say are the difficulties in a professorship?

I would say the challenges are the following:

- managing and leading a group, with the range of people dynamics it brings;
- there is always fighting with colleagues;
- there are administrative loads for sitting on committees, which as the only woman can become considerable. It is hard to find a balance on when to say "no" and when to agree;
- you have to distribute your time over a range of research topics;
- delegating to senior collaborators is normal, but can have disadvantages if they do not work well with you;
- it is a day and night job—7 days a week—and it is never possible to get through everything. You can feel constantly that you are not doing enough because there is always a tidal wave of things to do.

How do you manage to keep a perspective with such demands?

Every year I have good intentions because I do not want to lose the fun of science. I plan and take the time to make sure I have the freedom to do something that is really fun and is science and allows me to do what I want to do. From the beginning, I was convinced about carrying out computer simulations in Biology and I was told that what I planned was science fiction. However, I have done almost all that I envisaged and I am now focused on the next scientific questions that I think can be solved. Moving forward in science is, for me, more important than gaining more power and influence.

We have discussed a little your experience of being one of the few women in chemistry as student and before tenure. Did this have an impact on you as a professor?

There really are different experiences between men and women and the hardest thing is the unconscious bias. It is very hard to fight against it, and with time I became resigned to it being there. I feel it will not change in my lifetime. I had to learn to live with things that I cannot fight anymore. One recent example within a collaborative network is that other male professors tend to contact senior male persons in my group, rather than contacting me directly. It has happened many times in networks where I can be treated as invisible. I have learned not to waste my energy on worrying about this and, as a result, choose my collaborations carefully. When this happens I ask myself the questions: can I work with this person? Or should I move on?

I have also experienced the well-documented situation where a woman professor makes a comment in a meeting and is ignored, but when a male colleague repeats the same point a few minutes later he is praised. In addition, whenever there are nominations for prizes, or in job interviews, certain qualities in men are seen as a plus, while for women they would be negative. For example, a quiet woman is colorless and has no personality, while an extrovert woman is pushy, a troublemaker and aggressive. Another example is, I was the only woman on a selection committee where a man was described as full of hot-air and having little substance, yet he was put on the shortlist of candidates. I spoke up in that meeting about the discrepancies in language that were being used to describe men and women in this discussion. The Dean on the committee supported my opinion that this selection process was being affected by unconscious bias, which was important. The problem is, that if leadership says and does nothing about these biases, the process will not change. For example, I have witnessed many times in tenure discussions that the fact a man is married is used always as a factor to support his case. This automatic behavior is often not realized, even in the best of people.

What do you think you can do to make a difference?

Things are changing slowly in the right direction but I have had to ask myself "What can I change? What am I unable to influence?" Colleagues can switch off quickly on this topic. If there is a problem and I speak too much about the topic I can make it worse. The issues are deep and widespread and arise from the basic question: "What are the roles that men and women play in society?"

What can you do to inspire young people and women?

It's important for them to see female role models who are not superhuman, who talk honestly about the good experiences and the challenges. Women should know the limits in the career and learn from others to find ways to hang on, because it is worth it. In this job you are able address important scientific questions and to do this with the independence a professorship brings.

Can you highlight some of the positives in your career?

My two postdoctoral supervisors Mike Klein and Michele Parinello are really great and inspirational people. I am not very good at networking and not embedded in a powerful system, so I was absolutely delighted and surprised to be put forward for the Dirac Medal. Nicholas Handy nominated me, though we have had only met two times and exchanged papers. I was amazed to get the support for my research work from my scientific community, and spontaneously, because I was not being strategic about networking. As a young woman, I got too much attention, more than I'd wanted, so I did not socialize a lot as a result. Recently, I was nominated to the International Academy of Quantum Molecular Science, also unexpectedly, for my body of work. It was lovely to have this underlying sense of support from the community, and it was truly gratifying.

Prof. Clara Saraceno (Argentinian, American, Italian)

http://www.puls.rub.de
Photonics and Ultrafast Laser Science, Faculty of Electrical Engineering and Information Technology, Ruhr University Bochum, Bochum, Germany

Clara Saraceno, "Ruhr Bochum University, Germany"

Biography

Since 2016 Clara Saraceno is associate tenure-track professor of photonics and ultrafast science. She was born in 1983 in Argentina. In 2007 she completed a diploma in engineering and an MSc at the Institut d'Optique Graduate School & Ecole Polytechnique, Paris, France. She completed a PhD in physics at ETH Zürich in 2012. From 2013 to 2014, she worked as a postdoctoral fellow at the University of Neuchatel and ETH Zürich, followed by a postdoc position from 2015 to 2016 at ETH Zürich.

Research area

The RUB Photonics and Ultrafast Laser Science Group is interested in exploring paths to achieve high-power ultrafast coherent light sources in exotic regions of the electromagnetic spectrum. The vision of our group is to demonstrate unique light sources that allow us to better understand the dynamics of physical, chemical, and biological phenomena. The focus of our current research is the demonstration of compact and efficient Terahertz light sources to explore solvation dynamics, together with the Cluster of Excellence RESOLV.

Honors and awards

2018 ERC Starting Grant
2015 Sofia Kovalevskaja Award from the Alexander Humboldt Foundation
2013 QEOD Thesis Prize, Electronics and Optics Division of the European Physical Society
2013 ETH Medal

Conversation with Clara Saraceno—September 18, 2018

Your father is a physics professor, did that influence your choice of career?

There was actually no pressure on me or my sisters to study his field. We would go to his office and it was always fascinating and fun, but my parents put no pressure on our choices. In fact, my oldest sister became a ballet dancer and the middle sister is an interpreter/composer. My key influence came from my bilingual (French and Spanish) school in Buenos Aires.

I finished school with a dual high school diploma—a Baccalaureate for the French school system and an Argentinian high school certificate (bachillerato). This school was orientated to the children of French diplomats and most subjects, except some topics like history and Spanish literature were taught in French. The best students were pushed to take the math/science orientation of the high school qualification. I was a very good student, the top of my class in a lot of topics so I automatically took the science direction.

How was it possible for you to learn everything in French?

When I was 6 years old my father had a year's sabbatical in Paris and the whole family moved there with him. I entered the normal French school system for a year and was fluent in French at the end of it. When we returned to Buenos Aires my mother wanted us all to attend my bilingual school, because it was one of the best in the city. My older sisters had differing interests by then so it did not work out for them, but it was great for me. This school was also different because to complete both high school diplomas, the school hours were from 8 a.m. to 4 p.m. Most Argentinian schools are open only in the morning. I owe a lot to my school because it pushed me in the right direction. Another factor was that the school had very good science teachers, one was particularly inspiring, and his classes were fun. He died this year without knowing the influence he had on my life.

You moved to France for your undergraduate studies. How did this happen?

The students from my high school, with the best grades, could apply for scholarships to enter the Grande Ecole program in France. It meant entering a system, which involves 2 years of preparatory classes to take the competitive entrance exams, then subsequently entering one of the specialist Grand Ecole. I followed this program, spending 2 years in Lyon taking the "classe preparatoire" and then studied at the Institute D'Optique Graduate School in the suburbs of Paris.

The 2 years in Lyon were very intense, with 6 days a week classes, and continual exams. The exams were tough and you were always told that you were not good enough. That is the basis of this system. I made some lifelong friends during this time. At the end you are tested over a number of weeks in every discipline: science, literature, and languages. Afterwards you are classified as "admit or fail," depending on your marks. Even if you are classified as "admit" you need a certain level of marks to enter particular top schools. I aimed for the Institute d'Optique and was successful with my marks and the second stage of oral exams. I then attended this school/university on a scholarship for 3 years.

The whole process had a huge impact on me because of the negative education style, where you always got low grades and were never good enough, which I found demeaning and demotivating. The Grand Ecole system produces many people who go on run the country, industry and businesses. A good aspect of this system is that anyone can go through the process. It is based 100% on merit and not related to your family origins. I know people whose lives were changed by the system, but this is definitely not in the majority of cases.

What was your experience once you moved to the Institute d'Optique in Paris?

It's actually on a university campus area on the outskirts of Paris. I enjoyed the physics, electromagnetic, and optics courses, but the focus was really on mathematics and I felt a bit out of place because I was better working in the laboratories. One positive thing that came out of the overall experience was that I learned early not to be overwhelmed when I have an enormous amount to do. In the 2-year "classe preparatoire" it was impossible to get through all the work, pass the exams well, but somehow, I managed it. This experience helps me put things into perspective now.

Your next step was to move to California and work in industry? How did that happen?

During our masters course we had a visiting lecturer from Coherent, Europe, and he informed us about internships available for master's projects at their headquarters in Santa Clara, California. I was too late to get a masters internship, but there was also a scheme where they took "engineering trainees" who had just graduated. I really wanted to go and work at Coherent. Though I was in a committed relationship at the time, and my boyfriend was not keen on the long-distance, he supported me because he saw the enthusiasm shining in my eyes.

I actually worked in another French laser company, Teem Photonics, in Grenoble for my master's project work. I organized this placement so I could spend 6 months working in Grenoble where my boyfriend was studying for a PhD. We'd met in the early days in Lyon, and I then studied in Paris and he went to Grenoble, which meant that I had already done a lot of long-distance traveling to be with him at weekends. I really enjoyed my time at Teem Photonics, which makes among other things microchip lasers, and after the United States they even offered to sponsor me to do a PhD, but other offers led me to turn this down.

How would you describe your time at Coherent?

It was a great experience! I planned to be there for 6 months but stayed for a year. I could probably have stayed longer with them, but my relationship pulled me back to Europe. The best part of the time was that, as low paid trainees, we were given really exciting projects, which were not risky if they failed. We did not have commercial deadlines so we could have fun with experimentation. I found my own personal drive there. It was such a nice free environment I enjoyed working with the uncertainty and wanted to explore. I worked at weekends and learned to keep trying and trying again to find solutions. At this point I thought I would go straight into an industrial job when I returned to France, this was also seen as the most noble direction at that time, after an Ecole education.

I returned to France for my relationship without a specific career goal, though I knew I wanted to stay working with lasers, and planned to look for a job. However, I'd started already to hear that it could be useful to have a PhD in the laser industry, particularly if you wanted to become a team leader, which made think about taking a PhD position.

How did you then move to Zurich to do a PhD?

I actually heard about the Keller group from a scientist working in the next booth to me at Coherent in the United States. He was about to move to Time-Bandwidth Products (now Lumentum), in Zürich, a company established by Kurt Weingarten and Ursula Keller. I asked him about places to study in Zürich and he told me about the Keller group, the laser development research there and he also put me in touch with her. The Head of Coherent provided me with references too and I was invited to interview. I applied for PhD positions in Grenoble, Zürich, and Barcelona. In the end I fell in love with the Keller Lab and there was an opening in the Thin Disc team, and I really wanted that position once I'd visited. If had not been in a relationship I would not have hesitated to take the offer.

My boyfriend told me that he would not move to Zürich and he wanted to make a life in France. He warned me that he was not keen on the long-distance travel, but did not hold me back from moving. Then, a few months after my move to Zürich, he broke up with me. I was super in love with him and was devastated, we'd been together a long time. Now I can see that we did not fit. I like to travel, it is my family experience, while he wanted to stay in one country. I am now glad that he let me go, because I would have probably stayed with him at all costs!

What was your experience of doing a PhD?

It was brilliant. There is a lot of independence in the lab and people do the research because they like to do it; they have their own self-drive, freedom and space to try things out. It was great to work with motivated, ambitious and high-quality people. I enjoyed belonging to a research line with general goals, but having the flexibility to explore and find your own projects within a common goal.

When I started the PhD, it was not clear that I would be successful. In the first year and a half I got no results. Then I inherited projects from PhD students who finished, leaving experimental set-ups in a very good state. All the big parts and the system were in place and I got results. It was important for me, when I left the group, that I should leave a similar situation for the next PhD students, so I could pay back what I gained.

I got lots of pleasure out of doing the research. I'm the kind of person that if someone says something cannot be done, like "there is no way to get short pulses" and it becomes a common belief that it will not work—I always want to understand why and keep trying to find answers. In this case I kept trying different ways and, finally, it worked. It is important not to accept that all common beliefs are a given, you should try your own different ideas.

You met your husband in the research group?

The PhD experience started to improve on a personal level. I really enjoyed Zürich and the research group. We did lots of sports together, partied, went mountaineering, and generally had fun. My husband is a person who likes to do things well, make them beautiful and precise, but he is also open to bigger life-changing things like moving countries. We were able to do research work together and support each other. He is always relaxed and unjudgmental. It was great to work in the same group because we really understood each other's experiences. He is a lovely straightforward person open to all the different possibilities in life.

When did you think about becoming a professor?

It was never clear that I would go for it, though I was supported by my PhD supervisor to try. I always thought I had the option to move to industry, but I thought I would give the professorship route a shot. At the end of my PhD I was offered a really nice job a Trumpf Lasers, in Germany, but I decided to keep my options open by doing a postdoc. I had a good starting point, because I was nominated for two prizes for my PhD work, and I got them both. It is interesting that my PhD professor at that time, Ursula Keller, was pushing me to go to the United States for a postdoc at an elite University (I had an offer at Harvard), but this didn't feel like the right choice for me. I felt I was sitting on a goldmine project, which I had taken very far, and it felt wrong to stop there. I was already recognized by my community for this work. In the end, I took a more "unconventional" path (in spite of several negative comments) and decided to continue with a 50% postdoc in the same group, and 50% in another related project with the University of Neuchatel. Both places taught me many things, and I think this was the right choice for me.

What made you decide go for the academic path?

It was a series of events and possibilities that moved me in that direction. First, Prof. Martina Havenith, Director of Excellence Cluster RESOLV at Ruhr Bochum University, Germany became a visiting fellow at ETH Zürich under the FAST program. I attended her lecture series during my postdoc and found her lectures and the topic of water research very inspiring. At one point she said "Our research is limited by the power of our Terahertz sources." This led to a discussion between Martina Havenith and Ursula Keller where, afterwards, they suggested that I might move to Bochum to conduct research into the technical limitations identified. It was a potential first step to my research independence.

Underlying this idea was that I apply for a Sofia Kovalevskaja award, which gives young international researchers funding to set up their own groups in Germany. I was invited to the Ruhr Bochum University to give a seminar, visited the labs and the engineering department where I could be based. Martina and I worked on the application in the month before submission and my husband backed the idea of giving it shot. We had no idea what would happen.

I was totally uncertain about everything—whether I wanted to move, about choosing to live there, about moving countries. My aim was to give myself an opportunity, to keep the doors open, and to prevent myself blocking my options because of prejudicial thoughts. I realized that I could decide when I knew about the award.

Then you succeeded with getting the Sofia Kovalevskaja award. What happened next?

I got all sorts of reactions from people—congratulations of course, but some were negative asking me "Why do you want to move to Germany? Your salary will not be as good. It's not a famous place, etc..." I also received some great advice from a successful young professor who said "Your group will grow and be successful, regardless of your university's visibility, the point is the research results you achieve."

Initially I was not aware of how prestigious the Sofia Kovalevskaja award was, because it all happened so quickly. There were only six awards that year across all disciplines. I was considered an asset for Ruhr Bochum University, which was really a very positive start, and they were extremely welcoming in what they offered. At this point I was also unaware about the levels of professorial appointment, but was given great advice by Ruhr Bochum when I explained that I was moving there, with my husband and baby son. I was appointed as a W2 tenure-track professor, which was significant because, right from the start, by making me tenure-track, they were promising me the goal of a permanent position after undertaking a successful tenure process.

What has been your experience of working at Ruhr Bochum University?

I was appointed in the Department of Electrical Engineering and Information Technology and could not have had a better start. They are a brilliant, super-organized department which supported me through the appointment process. I applied as soon as possible for an ERC starting grant, indeed I was 7 months pregnant with my second son during the interview process. The first application round was not successful, but I did not give up and tried again the next year. The second time was successful and, as a reward, the university is putting me through the tenure process early. Ruhr Bochum University is very proactive and welcoming to young professors and they want us to stay and contribute to the university environment.

You decided to start a family through this time of change?

For us having a family was very important and we decided that we would deal with whatever came. We knew we could both get jobs in industry and support a growing family. It all happened very quickly. Our first son arrived at the same time as I received the Sofia Kovalevskaja award. Our second was born in Bochum. The move to Bochum was also facilitated by concrete dual-career support from the university, with my husband receiving a postdoc position for 2 years.

In addition, the support for parents with young children at Ruhr Bochum is brilliant. My first son was given a place immediately at the KITA, and when my second son was born, we set up a flexible childcare room on-site in our research building (ZEMOS). We organized this tailored to our needs, with other female professors in the same situation and support from the RESOLV cluster. This was run by two childcare professionals, and enabled me to leave my son for a few hours, work flexibly and still look after a small baby. This flexible childcare room is an innovation and we are working to make it a permanent.

RESOLV also gives exceptional support to new mothers by providing research support through dedicated staff positions. As a professor I have been awarded an extra postdoc position for 2 years, while a postdoc mother in the network is given the support of a PhD student.

How would you summarize your situation at this point?

I am still not convinced that I am cut out to be a professor; I think that they call this the imposter syndrome. I enjoy what I am doing very much and I enjoy the fluidity of the job. I think my main advice for an aspiring professor is that you need to seek out for yourself job opportunities and you need to try for them even when you are uncertain of all the outcomes. Don't listen to other peoples' prejudices and expectations, but find the answers for yourself.

Prof. Maria Schönbächler (Swiss)

http://www.isotope.ethz.ch/research/formation-solar-system-planets.html
Institute of Geochemistry and Petrology, Department of Earth Sciences, ETH Zürich

Maria Schönbächler, "ETH Zurich/Giulia Marthaler"

Biography

Maria Schönbächler has been an associate professor of isotopic geochemistry at ETH Zürich since 2012. She was lecturer and reader at the University of Manchester (2008–12). She held postdoc positions at Imperial College, London, United Kingdom (2004–05) and the Carnegie Institute, Washington, DC, United States. She completed her PhD at ETH Zürich in 2003.

Research area

Her primary research foci are: the origin and evolution of our solar system and the Earth based on meteorites and geological samples, including the Earth's earliest evolutional stages, such as accretion, core formation, the creation of the Moon and the development of the first continents. She also aims to determine the origin of the initial material, from which planets are built and to refine the chronology of events in the first million years of our solar system using radioactive isotopes. She is interested in the origin of volatile elements in the Earth and their history in the early solar system and develops new cutting-edge methods in isotope geochemistry needed to resolve important questions in the earth sciences.

Honors, awards, and positions

2016/2017	Head of Institute of Geochemistry and Petrology, ETH Zürich
Since 2017	Director of Studies at the Department of Earth Sciences, ETH Zürich
Since 2014	NCCR PlanetS—Leader Platform "Equal Opportunities" and Executive Board member

Since 2016	Executive board member of Swiss Academy of Science
Since 2018	Delegate Swiss Academies of Arts and Sciences
Since 2017	Chair Paul Niggli Foundation
Since 2013	Editorial Board, Geostandards and Geoanalytical Research
Since 2016	Editorial Advisory Board, Earth and Planetary Science Letters
2011	ERC Starting Grant "Accretion and Differentiation of Terrestrial Planets"

Conversation with Maria Schönbächler—August 9, 2017

Go your own way—Pave your own path

What made you want to become a professor?

My route to becoming a professor was, in the early stages, very untraditional. I come from a family with no links to academia, no one studied at university—like the majority of Swiss people at the time. I liked school, but had no interest in doing my homework, or making any effort to succeed academically. That was for the clever people, and I was just "normal." I decided to do an apprenticeship with the Swiss Post Office and I worked there for 3 years. I liked working with people, had a good interaction with colleagues and customers. I was happy with my life choices at first, but then grew more and more restless.

What brought about your change in thinking?

When I turned 19, I went to Brighton in the United Kingdom for 4 months to learn English. The time out made me realize, what was already becoming clear before, that my job was not challenging enough and that I wanted to return to study. Back in Switzerland I moved to a job in Lausanne, the French speaking part of Switzerland, to improve my French, and began to take distance learning courses to get the high school qualification, the Matura, required for university. For 2 years, I continued to work 100% at the Post Office in my permanent job and then, in the third year, I had courses on Saturdays, so I went down to 80%. This reduction of work time was not straightforward. My boss first told me that everybody likes to have a free Saturday, but that the post office is open on Saturday, so it would be impossible to reduce! I did not take this "no" for an answer. Sometimes it is worthwhile to be persistent and convince the boss of alternative solutions. It worked out for me. Once I got my Matura, I chose to go to University. This was a big step, because it meant I had to give up job security and to change my life completely. My hobby, which I used to do in the evenings and on weekends, became my major focus.

What speciality did you choose for your Matura?

It was actually the classics version with a concentration on Latin. I loved Latin, apart from learning the vocabulary. It also meant that I could study the sciences—physics, math, and chemistry. I liked the sciences as well and now that I'd reached the turning point I had to ask myself what I would like study. I knew that I was curiosity driven, and that this should guide my way. I chose geology and I gave up my job security, lived on my savings and worked at the post office in the holidays. Two weeks work at Christmas and working over the summer holidays enabled me to stay financially independent, though my family was also very supportive of my change of direction.

What made you choose geology?

I had always been interested in astronomy and planets, but the geology course at ETH had a wider curriculum, with more variety than the astronomy curriculum. I could study math, physics, biology and chemistry—all the sciences I loved—and go on geology field trips. Studying for a degree was enjoyable, but the math part in the first year was a wake-up call. I was used to learning on my own and math was so easy for the Matura, but at ETH you often had to go and find out the right information for yourself. This was new for me and I was not prepared for that. I did not know how to find the right books or who to talk to in order to understand the topic, or find the right documentation. Sometimes I hit a wall in this unknown territory! I had to find the way through it and learn how to find the right information. I also learnt at this point that much of the important information comes from networking. Coming from a non-academic background, I had not been aware of this. But what life had taught me before my ETH studies, and this is equally important, is that you have to be able to work with obstacles and not give in easily. You also have to like challenges and have the ability to persist.

In my third year of the Geology degree, I specialized in petrology and went to the Sierra Nevada in Spain to do mapping and geological studies. In 1995 there was about 15% women on the course, 10 out of 60 students. Nowadays it is at least 30%, and in the last 2 years it was for the first time even more than 50%. This diversity really changes the group dynamics in a positive way. I already knew during my diploma that I wanted to do a PhD, I liked the research parts and wanted to discover something new.

Was your PhD the beginning of your current field of research?

Yes, my topic was cosmochemistry and my supervisor was a new professor at ETH. I was drawn to the new research field, and the idea of stepping over the boundaries, and finding out about things that no one had discovered before. I was mapping out chondrites (stony meteorites) in areas where they had not been mapped. In Zürich, we were analyzing isotopes in these meteorites with high precision. It was a lot of work, but enjoyable and I completed my PhD in 3.6 years.

The next step was a postdoc position

I saw a job advert for a postdoc position at the Carnegie Institute in Washington DC and I applied for the job. The successful application meant that I could work on meteorites and a new isotopic system. It was also at this point that my dual-career issues started. Or maybe a better name is dual career challenge.

My partner did his PhD in cosmochemistry at ETH Zürich before I did, and I'd met him later, when he was already a postdoc at the University of Bern. After I was successful with my postdoc application to the Carnegie Institute in Washington, he wrote his own grant proposal, was successful and moved to a fellowship at the same institute. This pattern has repeated itself throughout our lives. I find the next position, and then he has found a way to pursue his career as well, either in the same place, or at a nearby research institute.

Negotiating the dual-career problem has been a considerable factor in your career, hasn't it?

Yes, let me explain. After 2 years in the United States, I moved to a postdoc position at Imperial College, London; this was to work with a former PhD supervisor from ETH, who was now at Imperial. My partner then found a position in Milton Keynes, at the Open University. They had just gotten a new NanoSIMS instrument, which is part of his expertise and he was offered a position related to that. We lived in Watford and commuted an hour each, in different directions. The next job offer for me came from the University of Manchester and was a permanent lecturer position. It was a great opportunity; they have a good cosmochemistry group there and I got my own lab and instrumentation. My partner then applied successfully for an Aurora STFC fellowship and, after a year of commuting back and forth from Milton Keynes, he also moved to Manchester. Later he would get a permanent position at Manchester too.

Then what lead you to move back to ETH Zürich?

I worked at Manchester University from 2008 to 2012 and was promoted to a reader position (just below professorship) during that time. Our dual-career situation had been solved and we both had independent permanent positions. However, I saw the advert for a professorship at ETH Zürich and I just had to apply for it. I did not expect to be successful. It was then great when I was offered a tenured permanent position, but there were no additional funds to solve the dual-career dilemma. I almost turned down the professorship, because this issue was not resolved, but my partner felt that he could not be responsible for me declining this unique opportunity. So, we gambled! In the end he gave up a lot! He had a permanent position at the University of Manchester and was then appointed as a senior scientist in my department—not a permanent position. I managed to find funding to support his position financially. He leads the nobel gas facility at ETH, which is his core expertise. It is a good position and, now that his old PhD supervisor has retired, he replaces that expertise, alongside leading his own research group.

What are the advantages for you to being at ETH Zürich?

I have access to all the university's great resources and now I can live close to my family. I was lucky that the position arose, due to a recently appointed professor leaving Switzerland for family reasons, after only 5 years at ETH. Cosmochemistry is a small field and in Manchester my group consisted of five people. I received an ERC starting grant at the same time as my application to ETH in 2011, so it was a very successful year. I now have a much bigger group, between 15 and 20 people. I needed to rethink my approach and work out how to progress with the research and supervise a bigger group, which needs different management than a smaller group.

At ETH I have more resources to carry out my research and I could develop my leadership and management skills. Three years after my arrival my group is pushing forward and we are excited by the open scientific questions, we can address, and the research possibilities.

Did you face challenges as a woman in your career?

Actually, being a woman scientist in the United States and United Kingdom was not a topic that arose. The overall fraction of women, where I worked, was generally >30%. In Switzerland, I did not have such an easy start, however. There were very few tenured women professors in the department at that time. This situation has improved and being a woman among many men mainly means being asked more often to address open tasks and positions. ETH gave me the chance to move to a different level in my career and this has been a significant and positive step.

In the meantime, I am part of the successful NCCR PlanetS and leading the platform for equal opportunity and advancement of woman. The aims of the platform include to support a new generation of talented woman researchers and help them succeed in science. It is interesting to observe that awareness and acceptance for gender specific topics is increasing and is becoming more and more accepted in the science community. However, it is still a long way before we fully recognize and account for the biases imprinted in our subconscious (Unterbewusstsein) that generally work against woman in science and leading positions.

What would you say were some of the highlights of your career so far?

I would say the experience of having my discoveries being widely acknowledged through the publication of my paper in Science in 2010. It was a wonderful moment of enlightenment when I was able to bring everything together from the data I'd been examining in Washington DC and to made sense of it. It was really satisfying.

Were there people who inspired and mentored you in your career?

My PhD supervisor, Alex Halliday, was a model for me in the way he handled being a group leader and how he managed the science. He knows how to motivate people using enthusiasm and persuasion. He showed me that it is important to have a vision, to make it happen, and then to advertise it to the world.

I am mainly very self-driven, though I am part of a community of researchers. My original career gave me the experience of working outside academia and how it is to have an 8 a.m. to 5 p.m. job with holidays. Academia takes up a lot of your life and you don't tend to have many hobbies. You need to have the dedication and passion, and this is the only way to make an academic career. I was always confident that science was right for me, but I also had to develop and grow the confidence to stand up, talk and teach. I had to learn to use my voice. I never dreamed that I would get here. This is another world from where I started.

Prof. Olga Sorkine-Hornung (Israeli)

http://igl.ethz.ch
Interactive Geometry Lab, Institute for Visual Computing, Department of Computer Science, ETH Zurich

Olga Sorkine-Hornung, "ETH Zurich/Giulia Marthaler"

Biography

Olga Sorkine-Hornung is a full professor of computer science at ETH Zurich since 2018. She was born in the Moscow region, Russia, and grew up in Tel Aviv, Israel. She leads the Interactive Geometry Lab and is the current Head of the Institute of Visual Computing. She joined ETH in 2011 as assistant professor and became associate professor in 2015. Prior to moving to Zurich, she was an assistant professor at the Courant Institute of Mathematical Sciences, New York University (2008–11). She earned a BSc in mathematics and computer science and PhD in computer science from Tel Aviv University (2000, 2006). Following her studies, she received the Alexander von Humboldt Foundation Fellowship and spent 2 years as a postdoc at the Technical University of Berlin.

Research area

Olga's research interests are in visual computing and computer graphics. Specifically, she works on theoretical foundations and practical algorithms for digital content creation tasks, such as 3D shape representation and modeling, processing of geometric data, digital fabrication and computer animation. The results of her team's research find their way into the making of films and video games, product prototyping and engineering software and medical devices.

Awards and honors

2017 Max Rossler Prize
2017 Eurographics Outstanding Technical Contributions Award
2013 Intel Early Career Faculty Award
2012 ERC Starting Grant (for full list see website above)

Conversations with Olga Sorkine-Hornung, April 29, 2015 and April 7, 2017

Where did your interest in math begin?

My family moved to Israel when the Soviet Union broke down, during the second wave of Jewish emigration in 1993. It was a culture shock. I went directly into middle school on my arrival and had to learn Hebrew by osmosis. I'd received a very strong mathematics education in Russia, and this part was easy for me, but in other subjects, such as history, I felt like an idiot because of my lack of language. At 12/13 years you change schools in Israel and I changed a year after my arrival. It was then, to my relief, that I realized I could understand almost everything.

You attended university very early, while you were still at school. How did this happen?

The school I attended was in Tel Aviv University neighborhood and it encouraged academic excellence. This really suited me because I loved studying. When Tel Aviv University set up a special pilot program for students from our school to take classes at the university, it was a great opportunity and I was deeply interested. You had to pass an exam to be accepted onto the program and when I passed, I was the only girl alongside four very nerdy boys. I attended first year undergraduate classes and sometimes had to miss school lessons, and when I left half way through a French class my teacher would say "Olga is going to teach at the University."

It was a strange experience at university in some ways, because all the students were significantly older. We were only 15 years old, while the other students had done their army service first—2 years for women and 3 years for men—so they could be 24 or 25 years old. The older students told us we were crazy and missing out because we did not have an ordinary school life. In contrast, I thought it was all truly wonderful, that everything was really interesting and challenging. I loved the experience and it was great to be saving time on studying.

I finished school and completed my undergraduate degree in Math and Computer Science at the same time, when I was 19 years old. I had not taken school holidays, but rather studied for my undergraduate courses in the break. I had also been allowed to postpone my army service a year so I could finish my degree first. I had no thought at all to become a professor at this point. I thought that my financial future would be in computer science, because of the internet boom.

You had to do 2-year army service. What was your experience like?

I'd hoped to be assigned a job in the army related to my knowledge of computers. At that point my career plan was already set: to become a cool programmer and earn lots of money for my family. My parents had broken their backs to provide me with an expensive neighborhood in Tel Aviv so I could attend the very best high school and university. It was not easy for them being immigrants. My mother's experience was very hard because she had trouble adjusting to the language. In Russia, she'd had a career as a mathematician and worked in a government institution, while in Israel she had to take temporary jobs, like looking after other peoples' children. My father had a PhD in physics, but there was a surplus of scientists from Russia, as almost one million people moved to Israel then. Jokes circulated about Russian professors sweeping the streets, and it was actually true. My father learned Hebrew, but first he worked in supermarkets and washed dishes at restaurants, before eventually finding an R&D job in a start-up company. Another factor was that my older sister stayed in Russia, which meant travel back and forth, and regular phone calls. It was a financial struggle.

You wanted the army service to continue the next steps of your career?

I wanted to use my computer science knowledge from my degree, but I discovered that to get a routine programming job you had to sign up for an extra 2 years' military service, and I did not want to prolong the experience. The alternative was to work in the intelligence corps. I would have really liked to do this but, because I was an immigrant and my sister still lived in Russia, I was deemed a security risk and was not allowed to sign up.

Instead I was given a clerk's job at the military headquarters in Tel Aviv, where I worked in customer service and had to answer questions about army benefits for officers. Being based in Tel Aviv meant that I had the advantage of sleeping at home. I worked for 6 months in this call center, but after a while I simply could not bear to spend all my time discussing various shades of nail polish with my coworkers. I called my favorite professor from Tel Aviv University, who had already given me specific computer graphics projects and asked if I could do another project with him. His positive reaction meant that I could explore the option of taking a masters at the same time as my army service. Given that I worked shifts, it was possible to fit in study sessions around my work.

My request to study a masters was very unusual, but my female army commander (who was only 1 year older than me) decided that, if I did not break the rules, and I studied in my own time, then I could enroll. It was fantastic to be back at the university, and the environment was so flexible that I was already doing research at 20 years old. In fact, my first research paper was published when I was still a soldier. My professor encouraged me to submit a paper to a conference at Manchester University, and it was accepted, and I was given special permission to leave the country to attend.

Did you already have plans for your next steps?

On leaving the army I wanted to finish the masters, go to industry and find a high paying job. When I asked my professor to help with a scholarship in order to finish my masters he agreed, but it meant applying for the PhD program to access funds. I caught the research virus and, when it was time to leave with only a masters, I could not imagine leaving without a PhD. I loved the research work; my PhD results were quite successful and my professor was an exceptional mentor. I was introduced to his network, visited the Max Planck Institute in Germany, traveled to various labs and conferences abroad, and was sucked naturally into the research world.

Your postdoc was at the Technical University of Berlin. How did this come about?

Like most of my career decisions when I was younger, the choice of location was motivated by my personal life rather than purely professional matters. Nowadays I realize that this is more typical for women, we tend to not dare to put the career first. I was in a serious relationship with a PhD student in Berlin and I wanted to give the relationship a chance. I applied for an Alexander von Humboldt Fellowship, which gave me my own funding, and a significant degree of independence with my research projects. I started my own projects without consulting my host professor, but we produced two extremely successful joint papers and he never complained that I did not meet his expectations.

The next career step also came from dual-career planning. My partner was a US/German citizen and we met in Germany where his father was stationed. He wanted to be a professor too and we were both in the field of visual computing. We realized that it would be impossible to have two careers in the same city in Europe, and that we had to move to the United States, because the big cities there often had several universities. I applied first, had five interviews, and received AP offers from Brown University and New York University. New York was the most exciting city and I decided to go there. A year later my partner got an AP offer from Rutgers University, which is an hour's commute from New York, and we both started as assistant professors at the same time.

How would you describe your experience as assistant professor?

Initially it was hell. You are unprepared for the enormous amount of multitasking required. You need to write grant applications, do research, teach and look after graduate students. It was overwhelming. I had been sheltered before, because I did not need to apply continually for funding. My ex-partner experienced the same process and we did our teaching preparation together, because we were teaching the same courses. However, an underlying career competition between us became more prominent, and we split up after 2 years as professors. NYU provided me with our apartment in central New York, so we had the awful problem that he had to find a new apartment in a city where the pressure on property is enormous. In the end, after I returned to Europe, he got a professorship in the engineering school of NYU, and things worked out for him on a lot of levels.

You moved to ETH Zurich as an assistant professor after 3 years at NYU?

Not long after this split I went to a conference and met a postdoc who worked at ETH Zurich, who is now my husband. We started dating and had a long-distance relationship between NY and Zurich. I'd already visited ETH's labs before we met, and around the same time the Head of the Computer Science Department contacted me about the plans for a new assistant professorship in computer graphics to be advertised within a year. The department's policy is to approach the top candidates across the world directly, and to create a list of candidates in advance.

I had a good publication record, but that there were also a list of extremely good people applying for the position. The professorship had two research positions attached, which was a gift of excellent resources. I prepared myself mentally to deal with the huge interview committee and did everything I could to prepare for the interview. Within 3–5 days I was offered the job, which was great.

Was there any hesitation in taking up the offer?

My (now) husband could have moved to the United States, but ETH Zurich is a much better school with great prospects. It is also closer to Israel and Russia. My husband did not want to become a professor and he has a research job at Disney (later he moved to Facebook). My vision was that I could work jointly with him on certain research projects. In general, there are a lot of job options for us here in Zurich.

Next you had to achieve tenure. What did this involve?

It involved my 3-year contract in the United States and working 3 years at ETH Zurich. The tenure evaluation process took about a year. Everyone told me it would be fine, but I could not relax until I'd achieved it. The tenure process was very new in the Computer Science Department of ETH Zurich. Only one person had managed it successfully before me and he was brilliant, while a couple of people were rejected. The current policy is to apply very high standards when recruiting and only hire people who are likely to get tenure, rather than appointing someone outside the profile and seeing if they can make it. Industry competes strongly with the Computer Science Department and it is very well paid, for example Google, Apple, and Facebook are strong competitors. Some professors with tenure have moved from the university to work in industry, lured by direct technology transfer and astronomic salaries.

Would you say that you had any obstacles in your career?

I think that, despite all my visible awards and success, I had a self-confidence issue looking back. When I applied to ETH Zurich I had to dare to do it. It was actually taking a risk to make the application, because I had previously felt too insecure to apply to the big schools, for example I did not apply to Stanford or MIT. I see this confidence issue with my female students now. I once approached a young woman to see if she was interested in a semester project and discovered that she was interested, but simply did not have the courage to ask. There are very few women professors in my field. I had the idea to find more female professors by identifying the women awarded ERC starting grants. Out of the 77 awardees for computer science only six were women, 8%, and most of them already had good positions in their host institutions, so we would have hard time recruiting them to our department.

We have a great culture in our department. The department head recommends that all staff check their statements to women, before they speak them. His advice is, if it does not work with male colleagues, do not say it, because it may be inappropriate. We have also created CSNOW, the Network of Women in Computer Science, where male and female PhDs and undergraduates carry out activities to help young women consider a career in computer science. Our "taster days" for girls have increased recruitment, with 70% of our female first year students having participated in these days.

What advice would you give to young women academics?

Firstly, learn to trust your gut. If your work is good, it will come out. Second, you have to learn to deal with failure. If your grant is rejected it is only 2–3 people rejecting you, not the whole community. Keep trying two, or even three, times. Finally, trust your own decisions. I once had a bad feeling about a postdoc candidate with great references and reputation. I overrode my intuition and, in the end, my initial judgment was right. You need to listen to yourself in order to improve your choices and to strengthen your decisions. You will become less influenced by external factors as you become more experienced.

Since we last talked you became a mother to twins, a boy and a girl. How did you manage this life-changing time?

The pregnancy went very well, though it was tiring and I had to take maternity leave earlier, because twins are a high-risk pregnancy. The children are both healthy but there were a lot of things I did not anticipate. I underestimated the impact of sleepless nights and having two babies at the same time. My husband was very supportive and had a flexible arrangement with his employer at Disney; as long as work was delivered, he could be at home a lot for the first 6 months and we cared for them together.

The biggest challenge came when they went into daycare at 8 months old. They immediately got sick, and then we both got sick, and it became a regular cycle. I'd expected everything to be back to normal, was committed to teach an undergraduate course on linear algebra (4 hours per week). I lost a lot of weight from the sicknesses, and was weakened by the pregnancy, so I had to take 3 weeks off work (over the Christmas period) to recover.

It took time to adjust on all levels, managing changes in my group, and readjusting my links to the research community. I had superb support from ETH to find a place for the twins in the university's excellent childcare facilities close to my workplace. It has been a wonderful experience to become a mother, and now life is back into a rhythm for my family, with everything proceeding well, also with my research.

In June 2017 Olga Sorkine-Hornung was awarded the prestigious Rössler Award (200,000CHF in research funds) for her innovative work in the field of computer graphics. In 2018 she was promoted to a full professorship.

Prof. Nicola Spaldin (British)

http://www.theory.mat.ethz.ch/
Materials Theory Group, Department of Materials, ETH Zürich

Nicola Spaldin

Biography

Nicola Spaldin is the professor of materials theory in the Department of Materials at ETH Zürich. She studied natural sciences at Cambridge University, where she obtained a BA in 1991. She then moved to the University of California, Berkeley, where she earned her PhD in Chemistry in 1996. She next worked as a postdoctoral researcher in the Applied Physics Department at Yale University, before moving back to California, where she was assistant professor (1997–2002), associate professor (2002–06) then full professor (2006–10) in University of California, Santa Barbara, Materials Department. She moved to ETH in 2011.

Research area

Research in the materials theory group uses a combination of first-principles and phenomenological theoretical techniques to study the fundamental physics of novel materials that are of potential technological importance. Projects combine the development of new theoretical methods, application of the methods to existing materials, design of new materials with specific functionalities and subsequent synthesis of the "designer materials." Specific materials classes of interest are: *Magnetoelectric Multiferroics*, which are materials that are simultaneously ferromagnetic and ferroelectric, and *Transition-Metal Oxides* with "strong correlations," in which the behavior of each electron explicitly influences that of the others.

Awards and honors

2018	Honorary Fellow, Churchill College Cambridge
2017	Fellow of the Royal Society
2017	L'Oreal-UNESCO For Women in Science Prize
2017	Materials Research Society, Mid-Career Award
2017	Lise Meitner Lectureship of the German and Austrian Physics Societies
2015	Koerber European Science Prize
2014–18	Thomson Reuters Highly Cited Researcher
2010	American Physical Society McGroddy Prize for New Materials

Conversation with Nicola Spaldin, December 8, 2015

Find your most interesting question[1]

Where did your choice to study science come from?

I was really good at science at school, in this situation it was normal or expected that you choose Medicine, but this was not for me, I was too squeamish. I studied biology, geology, physics, chemistry, and mathematics at "O" level, alongside literature and music. I love math, but I also enjoyed artistic topics. As none of my family went to university, it was important that I did not study abstract subjects, but rather those where I could answer the typical question "What job will you do after your studies?" My "A" level choices were physics, chemistry, mathematics, and music.

Why did you study music with the sciences?

I was given a clarinet at 11 years old, when I entered high school. That was the beginning. After that I took all the music exams and "A" level music was fun. However, I really loved the natural science subjects.

The next step was to study natural sciences at Cambridge University, which is unusual coming from a nonacademic family. How did this happen?

Ever since I was a small girl I wanted to go to Cambridge. I first went to a Catholic girl's school in Sunderland near where I lived. Then at 16 years old I went to boarding school, on an athletics scholarship. My parents were amateur athletes and they often hosted sports competitors and some teenagers stayed with us who attended a private school called Millfield in South West England. It gave me the idea to apply there for a scholarship, based on my own sport which was ski racing. There was a plastic ski slope in my area and I had also taken the opportunity to ski on snow in Scotland where I learned to deal with ice. I was interviewed for the lower sixth year at Millfield and offered a scholarship, depending on getting good "O" level results.

I had a fabulous time at this school. It was a boarding school and for me it was like going to a summer camp. The facilities were brilliant, the teachers excellent

[1] Nicola Spaldin, *Find your most interesting Question*, Working Life, Science, July 3, 2015, www.sciencemag.org.

and I thrived there. It also expanded my horizons as I met different types of people from all over the world. My roommate was the daughter of a CEO in Asia, lived in Singapore and flew into school on a private plane. Millfield made the transition to Cambridge easy. It had a special office to assist pupils with applying for university and helped us rehearse for the interviews. This ensured that a large percentage of the pupils went to Oxford and Cambridge. I chose to go to Churchill College essentially because of its modern facilities, like plumbing and central heating!

You took a gap year before you started university

Yes, I traveled first and then I worked for General Electric in London. Afterwards I continued to work for them in the holidays and this job was influential in my early links with industry.

When I arrived in Cambridge it was initially hard to adjust to studying again. The courses were great in natural sciences. I studied physics, math, geology, and chemistry in the first year and then specialized in the latter two topics. I found the physics environment unfriendly at that time, which contributed to me deciding not to specialize in physics. I knew I would not stay at Cambridge, I felt that Cambridge seemed a little too self-absorbed, smug and self-congratulatory. I also had no idea what a PhD was, or what it involved, until I spoke with the teaching assistants. At one point I even considered doing an industrial PhD sponsored by General Electric (GE), but decided against that, because I had to commit to work for GE afterwards.

You went instead to the United States?

I chose Berkeley because of my sporting interests. In my year out, I had transitioned from skiing to rock climbing and therefore made applications to Stanford and Berkeley, because of the climbing opportunities in Yosemite, and I applied to Boulder, Colorado because of the Rocky Mountains. At Berkeley you applied for the general chemistry program and then chose the specific research topic once you arrived. I chose to do a theoretical PhD and worked closely with an experimental group.

At this point, though I moved more to the theory side, I was still thinking of potential links to industrial research and practical applications. The field was semiconductor nanostructures. I thought I would go to Bell Labs afterwards to do research, but in 1996 they were closing groups down, and a lot of the people who I wanted to work with were leaving. My PhD project was really interesting as I was working on the structure of nanocrystals, which was a very timely topic. My PhD took 5 years, which is normal in the United States because the first year you have to take classes and there is a lot of teaching in labs afterwards.

You followed this PhD with a postdoc at Yale University

Yes, but I had already accepted a position as a tenure-track assistant professor (AP), on the condition that I could defer my start date and do this postdoc first. I'd started applying for AP positions as a PhD student. It was not unusual in material sciences and engineering to apply early, because universities realized that they needed to hire early so young scientists did not follow the natural path directly into industry. I applied to Bell Labs, the University of California, Santa Barbara and the College of William and Mary, Williamsburg, VA. I was pleased to accept the offer from UC Santa Barbara.

Hewlett Packard sponsored my postdoc and I lived in Silicon Valley first for a few months, working on aspects of magnetic hard-drives; but then they closed down that division. Once I moved to Yale the project enabled me to do first principles, ab initio work on electronic structures. I learned new basic techniques. My supervisor was Prof. Karen Rabe (who later moved to Rutgers University) and I still work with her today. During my postdoc period I could really focus on acquiring new skills and could be very productive without worrying about publications, as I already had the position at Santa Barbara. I was working in applied physics and could take risks, engage in knowledge development and acquire a new skill set. It was here that my interest in multiferroics was triggered by a postdoc colleague saying "It is a pity that there are no magnetic ferroelectrics, because we could collaborate." This was the start of a new research direction.

How did your move to Santa Barbara go?

I moved from Yale in the summer, and at first experienced some visa problems because of my British passport, but this was cleared up quickly and I was teaching already by the Autumn. The positive aspect of this tenure-track AP position was that I became independent from the beginning. The downside was the lack of financial support. I did have some start-up money, so I was not desperate, but I had to bring in my own funding. The tenure procedure assessed APs on their science and also the funding they attracted. That meant it was really stressful at the beginning, because it is always normal for your first round of grant applications to be turned down, but you have to get the next one, and this procedure—unless you achieve success—is difficult. After that you can evolve as you wish. I changed research direction as a result of my postdoc at Yale; I focused on the quest for multiferroics. The tenure process took 4–5 years, and as I established the field of multiferroics, I found success.

What led you to become a professor when industry always seemed an equally good option?

I think it was because during my PhD I found that I really liked teaching, up until then I expected to end up at Bell Labs rather than becoming a professor. The job of a professor is very busy and getting the tenure-track AP position at Santa Barbara was the first step. The tenure process there was transparent, we were given regular reviews and feedback. The big challenge is to bring in funding, once that happens it can really work. Another factor was their sabbatical process. I could be a visiting professor at other universities every 4 years.

Santa Barbara was a paradise and my colleagues were really great. The only problem was the University of California's lack of investment in science, education and engineering.

So, you were invited to join ETH Zürich?

Actually, I was asked by Prof. Ludwig Gauckler to suggest candidates for an upcoming position in the Material Sciences Department at ETH. It was a colleague of mine, Prof. Nava Setter from EPFL, who noticed that I had sent a list of candidates that did not include myself. She called me and asked why I was not interested. I realized that I was happy to visit ETH Zürich and discuss possibilities.

The facilities at ETH are just incredible, so are the support staff and the measures to ease a move. The head of the dual-career office is great. The Department of Electrical Engineering had already explored the idea of employing my professor husband before I came to interview. In the end he has a titular professor position in Information Technology and Electrical Engineering (ITET) and the university was really proactive in making it work and amenable to finding a good solution. It was a very positive move for us to ETH, which is really a paradise. It feels like the whole country enables ETH professors to be productive. We have great support staff and the students are well trained and respectful. Perhaps some people do not realize how lucky they are.

You have also played in musical ensembles since you came to ETH Zürich

I play clarinet in a community orchestra and in chamber music groups. It is a distraction from the science, but has also meant I have made many friends outside of the ETH through my interest in playing classical music.

Your experience has been almost fully positive. Would you say there were challenges?

A step that is very difficult is establishing oneself as a new assistant professor because of the quantity of work required and the range of skills needed. Being a professor is a demanding job, you are like a CEO of a company initially with only research skills. The current problem is that I simply do not have enough time to do all the things I would like to do at ETH!

Did you have key people in your career?

Yes, my postdoc supervisor Prof. Karin Rabe, Rutgers University, was a mentor and became a long-term collaborator. Prof. Ramamoorthy Ramesh, University of Berkeley, my long-term collaborator has also been very influential. He constructed a new field of science with me and also supported me for many prizes and awards. Finally, my school piano teacher is a memorable influence because she strongly advised me to study science rather than music!

Prof. Elsbeth Stern (German)

http://www.ifvll.ethz.ch/en/

Institute of Behavioral Sciences, Department of Humanities, Social and Political Sciences, ETH Zürich

Elsbeth Stern, "ETH Zurich/Giulia Marthaler"

Biography

Since October 2006 Elsbeth Stern has been full professor of empirical learning and instruction research at ETH Zurich and Head of the Institute of Behavioral Sciences in the Department of Humanities, Social and Political Sciences at the ETH Zurich. She is responsible for the pedagogical aspects of the ETH teacher training program. For the past 20 years, her work as a cognitive psychologist has focused on learning in science and mathematics. After her PhD in 1987 at the University of Hamburg she held positions at the Max Planck Institute for Psychological Research in Munich, the University of Leipzig, and the Max Planck Institute for Human Development in Berlin.

Research areas

The focal point of her scientific work is the acquisition, change and use of knowledge. One of her main themes is how knowledge transfer can be improved through the use of visuospatial cognitive tools. She has also investigated the interaction between intelligence and knowledge through experimental and large-scale studies.

Awards and honors

More than 100 of her essays have been published, many of which appear in high-ranking international journals. She is on the editorial board of a number of journals, including Science. Beyond scientific research, she is a well-known figure, giving interviews on current topics in education, including her criticism of the pseudo-science marketing of neuroscience. In 2018 she was awarded the Franz-E-Weinert prize of the German Psychological Society for excellent research and communication of science in the public interest. She is member of the Academy of Sciences and Literature of Mainz.

Conversation with Elsbeth Stern, October 27, 2015

What made you want to study science?

I come from a farming family in Germany, my parents were farmers. I had seen the difference in lifestyle between my relatives who studied and those who did not and I had already had intellectual goals at 15 years old. It was not clear that I would be allowed to go to gymnasium as a girl from the countryside, but I was very good at junior school and we had a good teacher who organized our school into learning groups related to the intellectual level of the pupils. Our ages were between 6 and 14 years in the school, and this system meant that even though I was a grade 1 level pupil, I was in grade 4 mathematics. In some of the villages 20% of pupils went to gymnasium and in others 0%.

My family valued education but more for boys than for girls. It took both my aunt and my teacher to convince the family that I should be able to go on to the gymnasium. I found the work quite easy there and studied with pupils from a similar social background.

Why did you choose to study psychology?

This happened because my natural science and math teachers were not as good as my literature and history teachers. Some of those were excellent. However, I wanted to study science, so psychology seemed to be a good compromise. I worked very hard to meet the high GPA necessary for psychology, which is a subject in high demand. I already knew then I wanted to be a professor. One time I voiced my ambition out loud, but the response was not very positive, so I kept it to myself. I knew what I wanted and realized that I had to be flexible about the path to achieve my goal. My school teachers had told me that there was no longer any room to make discoveries in physics and math, so for me it had to be another field, like psychology, because I wanted to make discoveries.

How did you choose the University of Marburg?

Actually, we did not choose. You had to get very good grades and then apply centrally in Germany for a university place. This central office chose the university for me. I would have liked to go to another university at that point. The competition for places in psychology in Germany was as high as in medicine, so I was extremely happy to get a place on a psychology course. Marburg was close to the place I grew up and, as it was among my listed options, I chose there. I really liked the way the course was taught at Marburg; the statistical education was especially good, which gave me a good base for my future research. However, I always wanted to leave Hessen, the countryside and study in a big city.

After my Vordiploma/Bachelors I moved to the University of Hamburg to finish my Diploma, leaving the countryside behind. Actually, the psychology program was not as good as Marburg and there were conflicts between professors. I really valued what a good base Marburg had given me. One professor was impressed by my statistics background, so I was given a student assistant position early on. I did my master's thesis in social psychology and I was always interested in a variety of topics and had a broad scope of research questions. I decided to remain in Hamburg for my PhD, with the supervisor from my masters, working this time in the field of diagnostics.

How did you make the bridge from postdoc to professor?

When I had finished my thesis, my supervisor had one open postdoc position. This was a very critical period for me because the university system had economic problems and I had a postdoc contract for only 1 year. I also had to move universities, because the rule in Hamburg was that you could only stay a total of 5 years. Three positions became available at the Max Planck Institute for psychological research in Munich and I got one of them. I was extremely lucky, more than I realized at the time.

You had to complete a Habilitation to become a professor. It was known in the field as "the license to stand in a queue for any professor positions that might become available," and it was likely there would be only six or seven open professors in psychology across Germany. I started to publish in international journals, which was unusual at that time, and my habilitation was based on the topic "learning math and science from a cognitive psychology perspective." I had a good mentor in Munich, Franz Weinert, who was the director of the Max Planck Institute. He supported everyone with good ideas, helped them to set the research agendas, and maintained peoples' passion for science. He was a wonderful person.

German reunification helped my career. There were a lot of academics in Western Germany looking for positions, and when the Eastern universities were rebuilt some new positions became available. I was appointed to a permanent professor position at the University of Leipzig. I realized that the research and scientific perspective there was a bit disappointing, but it was the first step as a permanent position. I was really glad that I'd achieved my aim, particularly being a woman. When applying for professorships in Germany, I had my first encounter with the "golden boy experience." I was astonished and learned that achievement did not count the way it should. Even though I was quite successful, and had published more widely, male colleagues without international publications were favored.

What was your experience like in Leipzig?

My colleagues were a lot of old German men who made sure all their friends got the professor positions, even though their publication rates were poor. I was very lucky when my position became available; there was no "golden boy" because it was a new field. I stayed there for 3 years and it was not easy with my colleagues. It was a stepping stone for the next position.

So how did you find the next place?

I applied to a number of positions and men, who did not have as good a publication record, got the jobs in front of me. I was known to be outspoken and critical of things that I did not like. A lot of real achievements were not acknowledged in selection processes and it's worth saying that several very good male colleagues suffered too. There was a strong "old boys network" which meant that they appointed their colleagues and friends, no matter how mediocre their output.

At that point I did not care. I had a permanent position and I had collected a lot of research data at the Max Planck Institute in Munich. The challenge was that at Leipzig I had a big teaching load and no financial resources. I started to pose the question "What about my research output?" and thought it might be the end of my research career. However, I was then offered the chance to move to the Max Planck Institute in Berlin for a permanent position, with the same salary and retaining my professor title. I moved to the new job for the research resources, the working conditions and the fact there was no teaching load. It was not a step back, though I moved out of the university system, and all the ugly experience was over. I had really good colleagues and could do my research. In addition, my then boyfriend, and now husband, lived in Berlin, so this was a personal improvement too, as we moved in together.

However, even in the Max Planck Society there was an "old boys network." I quickly became part of a committee, headed by the Nobel laureate Christiane Nüsslein-Vollhard to improve the situation for female scientists. At that time, they did not advertise leadership positions because they believed that they knew all the eligible people, who were men of course. Once you rose in academia you started to see there is a glass ceiling. In Munich, I had not seen it. My second time at the Max Planck I became more aware of the challenges, but at the same time the research opportunities were brilliant, my colleagues excellent and I would have stayed there for the rest of my life.

What triggered the move to ETH Zürich?

I was invited to apply for a professorship at ETH. There was no "golden boy" candidate and peoples' minds were open. This tendency to work with "old boys' networks" really has to be compensated by universities; they need to have policies to correct the weaknesses of human nature. The position at ETH had been advertised, but I did not see it. For me to move from Berlin it had to be a permanent job and with enough funding for my research. I also received offers from three other German and European universities at the same time. In addition, I had a dual-career situation, so this was also part of the negotiations with ETH.

My educational psychology work is very visible publicly and relates to policy, governmental and institutional change. I had a very high public profile before I came to ETH. In Germany, I fought against a professor who was using false neuroeducation and neuropedagogical arguments to change education in schools, even though he had no experience of the specific kind of learning at schools whatsoever. Germany was in crisis during that time because the results from the first Pisa study in 2001 showed the education system there was not doing well. The Max Planck Directors thanked me for my public commitment to stop these false moves in educational policy.

I have never been frightened of fighting on an issue. I believe that professors have to use their power to make a difference on important social issues. I take my interaction with the public very seriously and allocate the time to do it. If you are able to inform society based on your research, then it is the responsibility of a professor to this, and it is a serious part of your duties.

What do you think influenced your life choices?

I think you need to have the aptitude for research and you have to work very hard. I think that I was lucky that certain positions appeared when they did. I always had a positive outlook and somehow thought it would happen. I know some people who went through periods of unemployment before finding a professorship, but that did not happen to me.

How would you describe your job?

It is like being the manager of a smaller company, you end up with a big workload. The money is very good, but I manage a relatively big and heterogeneous group. In addition, I have been the Head of Department for the last 3 years. On top of that I have the political responsibility to the community. My husband and I decided that it was too difficult to manage a career and to have children. We made a clear decision that we could have a good life and we were aware that you cannot know what comes from having a child. We both felt this.

It has been a challenge that women are rewarded differently to men. There is a stereotype that women work hard and men are gifted. Another factor is that men still get the benefit of the doubt when there are mistakes. Our experience needs to become more straightforward. All successful women have experienced the situation in meetings where they contribute on a topic and there is silence, but if a man says exactly the same thing everyone agrees immediately. It is called "Hepeating." Every single senior woman has had this experience. The women do not get the benefit of the doubt.

My job is the best job in the world, but there can be times when there are so many things to do—for example when I was Dean. At these busy times, the passion for research can vary because I have huge responsibility. After 3 years as department head I have learned my passion for research again.

So how would you advise those who want to become a professor?

As a postdoc, you have to have an absolute passion for your research. You need to go with your goals and passion. You should take opportunities even if they are not perfect, because you don't know where they will lead. You have to think forward: try, apply for things, and take steps in a different place. In addition, you need to think broadly about research possibilities, look outwards and explore options. In the end your choices have to also be practical.

And how does your life work practically?

My husband and I have the same goals so we can work together and understand when renewal is needed and when you are exhausted. More men in academia want a work-life balance and would like to spend time with their children, or they have partners who have their own careers. It is good to see the widening of life goals within the professorial community.

Prof. Shana Sturla (American)

http://www.toxicology.ethz.ch/
Laboratory of Toxicology, Institute of Food, Nutrition and Health, in Department of Health, Science and Technology, ETH Zürich

Shana Sturla "ETH Zurich/Giulia Marthaler"

Biography

Shana Sturla is professor of toxicology in the ETH Department of Health Sciences and Technology (full professor since 2016, associate professor 2009–16). She was born in 1975 in Brooklyn, New York, United States. She completed a BS in chemistry in 1996 at the University of California at Berkeley, and a PhD in 2001 from the Massachusetts Institute of Technology (MIT). From 2001 to 2004, she was a postdoctoral fellow at the University of Minnesota Cancer Center, and from 2004 to 2009, she was an assistant professor at the University of Minnesota.

Research area

The ETH Laboratory of Toxicology is concerned with the chemical basis of mutagenesis and toxicity. Various projects at the interface of chemistry and biology aim to delineate relationships between chemical structures, enzyme-catalyzed chemical transformations, and cellular responses to environmental and dietary toxicants and cancer drugs. Particular interests lie in factors influencing susceptibility to DNA-alkylating agents and food-relevant exposures. Broad impacts of the knowledge and tools resulting from this work are intended to improve human health by impacting chemical risk assessment and disease prevention, as well as precision medicine strategies.

Honors

2014 Young Investigator in Toxicology of the American Chemical Society
2010 ERC Starting Grant
2007 American Association for Cancer Research Minority Scholar in Cancer Research
2004 US National Cancer Institute Career Development award

Conversation with Shana Sturla—August 25 and September 23, 2015

When did your interest in scientific topics start?

I became interested in science as soon as I started high school. I grew up in Brooklyn, New York, but when I was about 13 years old, and it was time to go to high school, my parents moved to the suburbs of New Jersey. Brooklyn was an ethnically diverse community and moving to the suburbs meant that we were the only Latino family in the new area. Maybe being in a school system with more resources for hands-on scientific experiences influenced my interest. As soon as I took a freshman biology class I was hooked and I took every single science class possible in the school: chemistry, physics and biology at basic and advanced levels. In the end, I ran out of the advanced classes so I went back and took earth sciences too, which was on a more basic track.

My father also influenced my interest in science. He helped me especially with physics and math classes, though sometimes he would work problems with me at a much higher level than we were doing in the class. At the time, it would annoy me that he made these difficult problems even more complicated, but it was of course a huge benefit. Another influential factor was that I tinkered with my father in the garage, so I learned to feel comfortable handling machinery. When I finished my PhD, I remember the person working in the machine shop, from whom I had borrowed numerous tools to maintain and fix my research equipment over the years, said to me, "You know how to handle tools better than any of the men in your lab."

What made you interested in getting a university education?

My father went from the Dominican Republic to the United States to study civil engineering. He met my mother in the United States; she had moved from the Dominican Republic when she was very young. I am an only child and my parents were very supportive of my success in school, it never occurred to me that I would not go to university. My parents gave me a lot of freedom. They were very supportive, but without giving me a lot of academic advice.

Where I ended up going to school was very influential on my scientific path, but also a little random. At the end of high school, I wanted to get away from New Jersey and had this romantic idea of going to California. I applied to many schools in California and decided to go to UC Berkeley. As an undergraduate in the United States you do not need to select a major right at the start, so I began by exploring the possibility of three topics: biology, chemistry, and environmental sciences. In fact, these are still the three topics that work together in my current field of toxicology. I liked chemistry the most and chose it as my major because I found it the most foundational to the other areas, and challenging to study. I was drawn to the challenge and felt that if I could master it I would have more possibilities. Berkeley has a strong Chemistry Department and among students it was considered to have a difficult curriculum, which made it attractive to me.

You seem to have finished your PhD very early, in your mid-twenties? How did that happen?

I was always young at school. My mother organized for me to start grade school a year early so I graduated high school and went to Berkeley at 17 years old. My undergraduate studies were 4 years, and my PhD time at MIT was 5 years, and I did this without a break. I had my PhD by 26 years old because I'd focused on school and did not take time out.

How did you choose to study for a PhD?

I was really passionate about research and as soon as I knew what a PhD was I wanted to do it.

Your PhD was at MIT. Was this your immediate choice?

As an undergraduate I did a lot of research to identify which lab I wanted to work in for my PhD and I got advice from PhD students at Berkeley. I chose a field of research and narrowed it down to individual professors in the field, where their science interested me. I followed this by applying for a PhD in the departments where those professors worked. It was not possible to apply directly to them, because in the United States you apply to the department. I applied to MIT, UC Irvine, Caltech and the University of Colorado. Then I attended recruiting events and spoke with the individual professors. Following that I wrote to my favorite professor asking whether I could work with him if I chose to go to that university.

But how did you know as an undergraduate the research areas that interested you?

I had a lot of research experience in chemistry as an undergraduate student. First, I did a federally funded work-study experience at Lawrence Berkeley National Lab with Dr. Richard Fish. Then I took a practical course in lab work and taking this lab course lead me to my second undergraduate research experience. A female teaching assistant for the course recommended me to the professor as the best student on the course, which meant that I was given the opportunity to do research in the second lab. It was a turning point for me, being invited for this position with Prof. Clayton Heathcock, what I'd call a really critical moment. That PhD student was Anna Mapp, who is now a professor at the University of Michigan and she is one of my scientific heroes. All of my early lab experience meant I was able to choose the research topic for my PhD quite specifically.

How would you describe your PhD experience?

I have just come back from an event in the United States to honor my PhD adviser, Prof. Stephen Buchwald, on his sixtieth birthday. I learned a lot at MIT. I remember my PhD adviser suggested that I apply for a fellowship. Actually, I remember he asked for my CV in order to put in the application, and my response was "What is a CV?." I was pleased to inform him that I had already been awarded a fellowship for minority PhD candidates, so this was a great start. Initially I was interested in organometallic chemistry and this was the topic for my PhD. However, when I worked purely on methods to synthesize chemicals, I missed the connections to more applied research, and also the links to my interest in the environment and health.

At MIT there were a lot of toxicology labs and I started to attend their seminars and took a biochemistry class with one of the professors, John Essigmann. I asked John about doing a postdoc in the field. His first question was "Do you have a significant other?" I mentioned that my boyfriend of 1 year had just moved to Minnesota, but I was open to moving anywhere in the United States. He recommended that, by chance, there was a great professor at Minnesota in the area of toxicology. He then connected me to Prof. Stephen Hecht and I showed up in his office in Minnesota and told him that I would write applications for fellowships so I could work in his lab. During the application period I also interacted with him further, and this experience was so positive, that he offered me a position without the need to bring in funding. In the end, I was successful with both grant applications I submitted, one from the National Institute of Health and the other from the American Cancer Funding Society.

How did you take the next step to assistant professor (AP)?

After having worked as a postdoc in Minnesota for 2 years in the Cancer Center an AP position came available in Medicinal Chemistry. It seemed to be a little too early to apply for this position. However, at the same time, my relationship with my boyfriend was now serious, which meant that we were motivated to find jobs in the same place and he already had an AP position at the University of Minnesota.

It was an early job opening and an ideal position. I had publications from my postdoc, but not yet as first author, but I guess I had good recommendations. My PhD mentor advised me not to apply anywhere else at this point, given that it was very early it would not be positive to have rejections that would be remembered if I applied again the next year. I think that this was really good advice, not to apply too early elsewhere. The great news was that I got the job, and I asked to extend my postdoc period another year in order get more experience, which would benefit my research. Thus, I was able to develop my knowledge in a new area, finish my projects, publish my results, and also apply for grants that I could already have running very quickly after starting as an AP.

How would you describe the AP experience?

It was a tenure-track position and a great place to be an AP. I had good career development advice and support. There is a very well organized procedure for developing APs at the University of Minnesota. Historically the university may have been poor on handling tenure, particularly for women, so as a result of problems with difficult cases, they were forced to develop a very good and transparent tenure procedure, which made it clear what you had to do to be successful. I was an AP there for 6 years and was very clear of where I was in my career. Our excellent head of department was proactive about retaining and supporting the faculty and she approached me about applying for tenure a year early, which gave a very positive signal that I appreciated.

This was just before your move to ETH Zürich, wasn't it?

At this point my husband, who is 5 years older than me had tenure. We both have basic backgrounds in Organic Chemistry, but his current area is Environmental Chemistry. In 2008, he was approached to consider applying for an open position at the ETH Zürich. At first, he was not so interested, but after he visited ETH he became more interested and was offered a full professorship in December 2008. In January 2009, we visited ETH to negotiate with the ETH President. I was the accompanying spouse, but I had already investigated to see if there were any possibilities at ETH in my field. I found out that an AP position in toxicology was about to be advertised in the Institute of Food, Nutrition, and Health. I submitted my application on the day it was advertised and the process moved forward quickly. My new colleagues were welcoming and genuinely interested in my research and I integrated well within the institute bringing my expertise in toxicology as a new topic in the institute.

Was your appointment at ETH as a tenured professor?

It was vital to have tenure in order to agree to move. My husband would not have applied to ETH Zürich without the invitation and we were both very happy in Minnesota. I was close to tenure and had very good signals from my department. The ETH also evaluated my case by obtaining external letters, and I was appointed as an associate professor with tenure. Five of my group came with me which meant we were able to set up the lab and get everything in place in 2010. The construction people at ETH were good, made plans for my labs and did a very detailed job to put all the important technical needs in place. It was a very positive experience overall. However, I was a bit shocked when I walked into the first planning meeting for the renovation project and it was entirely in German!

Another positive factor was that Prof. Laura Nystrom was appointed in my institute at the same time as tenure-track AP. We were connected by ETH during the appointment process and really hit it off. We worked together as we set up our labs, it was a great way to start. Laura already knew some German so she could help at the construction meetings. As she was also starting from scratch we initially shared some common space, including group meeting rooms, though we separated once both groups grew and more space became available. It was a really good start and Laura and I have a very close relationship.

You started a family after you came to ETH

We began to discuss having a family before we left the United States. My husband was open to starting earlier, but it was too stressful to try during the move and after a year in Zürich I was ready. It helped enormously that I had people from Minnesota around me and my best friend worked for a couple of years in my husband's group, which meant I had incredible support. I had a fabulous pregnancy and was in great shape. The Swiss Health care system is simply wonderful. The dual-career office helped find daycare for my son. It helped that we were able to be flexible and wait until most places came available in August. My mother came for 2.5 months from the United States, and then my cousin, who was a college student, so my son went to daycare at 8 months old.

Would you say that you've had any challenges in your academic career?

It's interesting—my husband says that my sporting background has really helped me to make quick decisions in stressful situations and to deal with challenges.

What is your sporting background and how did you get into sport?

I was an asthmatic child and the doctor said I should get into a sport. I was swimming competitively by around age 11. In high school I was on the cross-country team, the track team, the swimming team and the diving team. When I went to Berkeley, I was approached by recruiters for the crew team and got into rowing. I joined the crew team when Title 9 was being implemented meaning that institutions receiving federal grants, like UC Berkeley, had to equally fund men's and women's athletic programs. There was already a great tradition of women's rowing at Berkeley but now we had salaried coaches, a travel budget, good equipment, support staff and we moved into the men's boathouse. I was a real beneficiary of the Title 9 investment, even though it was very controversial at that time. I rowed at Berkeley from 1992 to 1996. Then I moved to Boston and was rowing with a club and got into doing triathlons, which lead me to realize my passion for cycling. I started road bike racing around 1999 and from 2003 to 2008, I was racing at an elite level with a team—so pretty serious.

Returning to challenges then, are there any particular ones that you would identify?

Yes! I had a very difficult time when I first started teaching as an AP in Minnesota. I taught my first course jointly with a senior male professor who would sit in the back of the auditorium with a microphone and, during pauses in the class, would interrupt and inform the whole class what I had done wrong. The first semester was incredibly difficult. In the second semester, with the help of a senior colleague, I told him he could not come into my class. In the third semester, I was able to say "you take this part of the curriculum and I will take that." It was a very stressful time over a 3-year period. He was older, established, close to me scientifically but did not encourage. The situation changed with the appointment of a new head of department who came in and immediately took leadership in this process. She was strong, fair and professional, and took the problem seriously. She counteracted his behavior and the subsequent 3 years were much better. However, this experience had a long-term impact on my confidence in teaching and made me question whether teaching was my strength.

A second challenge comes from the review process for publications and grants, particularly if you are doing interdisciplinary work. There are growing pains in science because of this, as more research directions become interdisciplinary. A number of reviewers are specialists and do not have experience of the interdisciplinary science that is being funded. I can handle feedback on grant applications or papers if it is based on facts, but when one influential, anonymous reviewer says simply "I don't think this would be of broad interest" and their statement results in a rejection, even if the other reviewers are positive, I find this very frustrating. A professor then has to stay really positive in the whole review system for publications and papers, for the sake of their PhD students and postdocs, whose career depends on this process being successful. It is absolutely vital that you develop resilience and stamina with the publication and grant process.

Finally, what advice would you give to young scientists who would like an academic career?

To really commit yourself to strive for that academic career. I see many people hold themselves back because it seems hard, or there is risk, or they lack confidence. If you go for it, but it does not work out, that's ok, there will be other possibilities along the way and you can choose a new path. But once you really hesitate, the opportunity may well be lost. It's just like bike racing.

Prof. Effy Vayena (Greek/Swiss)

http://www.bioethics.ethz.ch/

Bioethics, Institute of Translational Medicine, in Department of Health, Sciences and Technology, ETH Zürich

Effy Vayena,"ETH Zurich/Giulia Marthaler"

Biography

Effy Vayena is full professor of bioethics in the ETH Department of Health Sciences and Technology since 2017. She was born in Greece in 1972. She studied history, history and philosophy of medicine, bioethics and health policy. She received her PhD from the University of Minnesota and completed her habilitation at the University of Zurich. Before returning to academia, she spent the early years of her career as technical officer at the World Health Organization in Geneva.

Research area

Vayena's research focuses on ethical and policy challenges in precision medicine and digital health. The advances in genomics, omics and data intensive research, have brought medicine and healthcare to an inflection point. Such progress holds great promise for improving individual and population health, while also presenting unprecedented ethical and regulatory challenges in areas such as privacy, individual freedom and risk policy. In order to harness the full potential of those novel approaches in healthcare, robust ethical policies are needed to resolve novel and pressing issues.

Honors

2015 Elected member of the Swiss Academy of Medical Sciences
2015 SNSF Professorship Award

Conversation with Effy Vayena, October 15, 2018

Where did you start your journey to becoming an academic?

I come from a small island in Western Greece, Lefkada, which was relatively unknown and fairly isolated at the time. But there used to be a lot of interest in art, poetry, literature and education. Reasoning, thinking and creativity were vital in the environment where I grew up and the attitudes were very open. I went to a public school on the island where the standard of teaching was high. You cannot be anonymous on an island of this size, and it was important to be a good teacher there. My parents were keen that my sister and I read widely, and their goals for us were a good education and for us to become economically independent. This happened; my younger sister has a PhD in mechanical engineering and a high-level job in the industry.

How did you choose to study history at the university?

Actually, I could have chosen many topics, because I loved studying and I particularly loved math, physics and chemistry, the natural sciences, literature and languages. However, around the time when we had to choose out specialisms at school my mother became very sick and she died when I was sixteen. I had to find a way through this very difficult period and it was a bit of a blur. During this time, I chose to take the classics route and to study ancient Greek, Latin, and history. One of my teachers was very upset that I did not study the natural sciences, but studying the logical structure of ancient languages was almost mathematical, and I loved that part.

I did not enjoy my history degree and finished it as soon as I could. It wasn't possible to change your major, but I discovered some courses on the history and philosophy of science in the engineering school at the Technical University of Athens. I attended these courses, even though I could get no credit points, because they were at another university. This decision to join these courses influenced my next steps.

You moved to London to study for a masters. Why did you choose there?

I applied to study a well-known masters in history and philosophy of science at University College London. I wanted to know more about this topic. I loved the course, it was totally open-minded with excellent professors teaching the material, such as Prof. Janet Browne, now at Harvard, Lara Marks now at Kings College London and Prof. Roy Porter (1946–2002). My master's dissertation was on In-vitro Fertilization (IVF) and assisted reproductive technologies. This year was the catalyst for the direction I would take in my career. I was asking questions about the impact of new reproductive technologies on society, and exploring the positives and downsides of the new developments. I was fascinated by the ethical challenges involved.

Why was your next step to the United States?

I always wanted to study in the United States, it was a given in my life, and I applied to universities in the United States to study for a PhD on the topic of History of Medicine, Science and Technology. My choice was the University of Minnesota, which also has a Center for Bioethics. I was awarded a fellowship for the PhD program, which made the move possible financially.

Initially I felt like I'd landed in the middle of nowhere, and it was extremely cold for a large part of the year. I was only 23 years old, and I had to be brave and find my way. I was very enthusiastic to be on a path to research topics related to ethics, philosophy, science and history. The program was really good, I made great connections, and everyone was really friendly and open. It was amazing to be in a university where it was acceptable that you can mix ideas across disciplines; it was so different to Greece.

What made the environment so amazing was that the professors, whatever their seniority, were open to discussions with a young person such as myself. I could visit other departments and meet with full professors who would spend time with me to discuss my topics and ideas. My PhD focused on the introduction of cervical screening programs, the development of the science, the introduction of the technology into a healthcare system and how it fitted with issues and ethics of public health. There was also the dimension of social equity: who could get access to the cervical screening, the rich countries only, or would it be extended to the lower income countries? This topic raised ethical issues in terms of clinical practice, public health and resource allocation. I flourished at the University of Minnesota with this open environment where distinguished professors gave me time and advice. I won a dissertation grant and spent the last year of my doctorate at Harvard University, where I had the opportunity further to interact with great thinkers and scholars.

Your next position was with the World Health Organization (WHO)? This seems like an unusual move

I knew before I finished my PhD that I wanted to do a postdoc and started to ask around for possible next step positions. Then I had a conversation with Prof. Cynthia Myntti (Emeritus Professor of Public Health Practice, American University of Beirut) who was actually on the Advisory Committee for my PhD. This conversation influenced the next steps of my life. Cynthia has this belief that advising younger people and women is of huge value, even though she receives no personal gain. She said "Why not work for the World Health Organization, because they are looking for people with your mix of skills?" It became clear that you could do research at WHO and she suggested I send them an email enquiry. She also wrote an excellent letter of reference. As a result, I had my PhD defense in November, an interview with WHO in their Geneva Headquarters in the following January, and I began working there already in February. My life in Switzerland began.

You worked for WHO for 7 years

Yes, but that was not my plan. I expected to stay only a year and move on. WHO had two research departments, one working on sexual and reproductive health and the other on tropical diseases. I joined the first section working on research on the ethics of IVF and reproductive technologies. We did research that linked together ethics, policy, medical technologies and public health. I had never imagined that there could be a position like this. I had the opportunity to teach the ethics of research in a broad range of countries, I was part of research projects and involved in the development of health ethics guidelines used across the world.

It was a wonderful opportunity. As an interdisciplinary person I could pave my own way, shape the functions around my expertise and bring together different strands of research. It was an incredible learning opportunity because I was working with the research world, the policy world, and with the people on the ground in different countries. You experience the values of linking pragmatism to research direction in these circumstances. You also learn some lessons about being in the public eye when representing an organization such as WHO. I had the opportunity to observe this environment. It was like a global school, watching how important advisers from all over the world operated, including the responses to the experiences of women and health.

Why, then, did you leave such an interesting position?

There were actually two reasons for leaving, both important. First, the one disadvantage of WHO was that I could not come up with my own research project, and I had the urge to develop my ideas and work on my own. The second, very powerful reason, was that my first daughter arrived and I could not balance the travel commitments that were the nature of this position with looking after my daughter while her father had an inflexible job in the healthcare sector.

Did you have a dual-career situation as well?

After the birth of my daughter I was already living in the German side of Switzerland, where my partner worked, and commuting to Geneva daily. It became untenable. Another factor was that the general culture was very challenging if you wanted to be a working mother. To pursue your career you have to be very strong and overcome a lot of judgemental behavior and opinions. You need a lot of mental strength, because you get asked often "You have kids, how can you have a career?" All the parental responsibility is still put onto the women and one very difficult thing is that, if you are excluded from the network of mothers, because you are working, your children get excluded too and pay a price. There are lots of subtle hints that you are doing things wrong. I think it is absolutely vital that women should have all choices in life: to be a mother who stays at home, or to be a mother who works part-time for a while, or a mother who works full-time for her career. It should be possible to do this without judgement. It is vital for young women and young parents to have options.

Young women look up to us female professors and it is important for them to see a range of female professors who have made different choices. We have to give young women the opportunity to be themselves and perhaps be delayed in their careers, but still be able do it later. One of my biggest challenges was being a mother, and wanting to pursue a career. At the moment legal changes, which make childcare available, are happening faster than the cultural changes, so there is definitely wider childcare provision than before.

Your next position was as research fellow at the University of Zürich?

We moved to Zürich as a family. Then I worked as a research fellow on short-term contracts. In this career route, you are left to find your own way, there is no recipe for success and it can be energy consuming, because you have to be determined and resilient. You have to hold onto the research and make the output happen.

What changed things for me was that I decided to apply for an SNSF professorship, even though I was older than the average applicant. I knew that the quality of my research output was of high value and I decided to approach the SNSF to explain why I was late in making my application. The SNSF allowed me to apply for the professorship and this was a turning point in my career. The SNSF did not punish me for taking time to deal with family matters, though they could have rejected my application on technical grounds (my age). I was honored that they allowed me to proceed, it was fundamentally important to me that they did this and I am very grateful.

Once I was allowed to apply I was confident about the assessment process. I had made sure that I had received my habilitation while I worked as a research fellow, and I met all the research criteria. Becoming an SNSF professor in 2015 opened doors from Day 1. I moved institutes at University of Zürich to establish my new group. This is another positive aspect of these professorships, with a successful application you have the independence to take your funding wherever you choose. I found a friendly intellectual home at the Institute of Epidemiology Biostatistics and Prevention led by Prof. Milo Puhan.

How then did you become a full professor at ETH Zürich in 2017?

My research topic of bioethics is a growing contemporary field and I received two written offers of permanent professorships, once I had the SNSF position. In 2017 ETH Zurich opened a new medical program, and the topic of Bioethics is a core topic for this new initiative. I was approached by ETH to see if I would be interested in my current position. ETH is a great place to work and I am proud to have joined this university. I feel honored to be here.

What factors would you say have influenced or been vital to your career?

I have been lucky in my career and came across incredibly talented and wise people who supported and mentored me. Bishop Hatzinikolaou from my college years, Dr. Marks, from my masters years, Prof. Myntti, to whom I owe the opportunity to work for WHO, Professor Kahn now at John Hopkins University, and previously at the University of Minnesota, with whom I still collaborate, Prof. Hafen who helped me navigate the Swiss academic system and who is a great example of resilience, and many more. They come from different paths of life, countries and academic disciplines. What they all have in common is a generosity of spirit and kindness. They offered support and advice when I was an aspiring young academic, (and even when I was a not-so-young academic)- this made a huge difference to my career and to my attitude. I want to perpetuate this attitude of being supportive and kind to others. I would love to reverse this tendency that becoming a tough person is the only way to go professionally; being nurturing, and kind and generous is the better way. It is certainly the more rewarding and meaningful way in my view.

I feel I was absolutely lucky to have had the funding opportunities that I did, starting with the PhD fellowship in the United States and then of course being eligible for the scheme of the SNSF professorship. This professorship was truly a career maker for me.

What would be your advice to young women who would like to be a professor?

Don't give up, stick at it, the "it" being finding who you are and why you want to follow this kind of career. Try to transform the tough times into what you have learned from them. There will be a space in the world where you can forge your own direction; you should not be overwhelmed by how others found their way, just find yours. There is space in which you can be you. Your talents and your weaknesses make you human. You don't need to abandon who you are to succeed. Improve, be better, but always be true to yourself. I am a poetry lover, so to all people and most certainly to young people I quote Mary Oliver "What is it you plan to do with your one wild and precious life?" This is a good question the answer to which will help us all with our life objectives.

I also think it's important that we don't just rely on simply telling young women that they need to be resilient to succeed. Being tough and hard-skinned is not the kind of virtue that we should promote. I have come to think that this can be a cop out. Women have to be extra resilient and tough because they face challenges about which we can do something. We can make these challenges more humane, so normal resilience would suffice. We should change society and academia, so that the barriers are reduced and women don't have to be superhuman to succeed.

We need plurality in our role models for young people. I firmly believe that it doesn't work to give women one role model of academic life that they should follow. Women should succeed by being themselves, whether they take time to have a family, or they have different sexuality, or they like to dress in a feminine way, or in a less feminine way. Academia needs to accept people for their academic talents and merits. So, when I tell young women to stay on track, I also tell me and the rest in academia that we have to do our job and change things.

On the challenges of academic life?

I do think that academia is still geared toward men, which influences the low numbers of women professors. In many fields of study we have over 50% of female students, we have a high percentage of doctoral students. Yet, we still have low percentages of female professors, Nobel laureates, university presidents, or other high-ranking academic officials. Implicit bias is a big problem. I am not convinced we do all we ought to be doing to change this. At the same time, I do know that changing one thing isn't going to change everything. We need more systemic change, and we in academia have an obligation to be part of the change if not to serve as a model for that kind of change.

The topic of sexual harassment is very current at the moment and I realize that I was always super vigilant on this and walked away before there was any possibility of untoward relationships. Being a woman anywhere, but surely in the workplace means that you have to be vigilant—and this is again something that tells you a lot about the extra worries of women. We should offer working environments in which women feel comfortable and not threatened, an environment in which they do not have to worry about more things than their male counterparts. I am vocal about zero tolerance to any form of sexual harassment and intimidation. I do this for my own sake, but also for the younger women who may not be able to react or respond to incidents of intimidation within asymmetric power structures.

To close this on a positive note: I find academic life to be a privilege. Of course, like all other areas of life, it has its issues and we should keep working to address them, but academic life offers you the privilege to remain curious, to investigate, to work on reason, to share knowledge with others, and to have impact on young people's lives and careers.

Prof. Brigitte von Rechenberg (Swiss)

http://www.cabmm.uzh.ch/en.html
Musculoskeletal Research Unit (MSRU), Center for Applied Biotechnology and Molecular Medicine

Brigitte von Rechenberg

Biography

Brigitte von Rechenberg is professor of veterinary surgery (since 2007), at the Vetsuisse Faculty, University of Zürich. She was the first female Dean of the VetSuisse Faculty, from 2014 to 2018. She was born in Switzerland. She completed her diploma in veterinary medicine at the University of Zürich in 1978. She spent 2 years at the University of Philadelphia as an intern and research fellow. Then, from 1980 to 1991, she established and ran a clinic for small animals specializing in orthopedics and trauma with her then husband. Returning to the University of Zürich she gained her Habilitation in 1999, and established the Musculoskeletal Research Unit. In 2008 she worked to create the successful Competence Center of Applied Biotechnology and Molecular Medicine (CABMM) together with her colleagues S.P. Hoerstrup and M.O. Hottiger.

Research interests

Brigitte von Rechenberg's main research interest is bone and cartilage research focusing on mechanisms of cartilage remodeling, resurfacing and substitutes for cartilage and bone. She completed and published several research projects, where osteochondral plugs were used as cartilage replacement. In addition, she worked on projects related to the spine, such as vertebroplasty, disc degeneration and posterolateral fusion. She researched the use of various biomaterials, for bone and cartilage substitutes, in conjunction with bone enhancing factors as well as basic mechanisms of cell signaling and interaction between bone resorption/formation. She is an expert in bone histology and her laboratory specializes in bone and cartilage histology using methods with nondecalcified specimens, routine paraffin histology and immunohistochemistry.

Awards and honors

2008–18 Head of the Steering Committee of the CABMM
2007 Charles S. Neer Award
2005 KTI Medtech Award
2004 Recognition Prize of the AO Technical Commission

Conversation with Brigitte von Rechenberg, July 1, 2015

When did your aspirations to study veterinary medicine begin?

I always wanted to be vet, as early as 3 years old. My godmother's husband had a veterinary practice and I adored the place, loved animals and I adored her husband, Uncle Christmartin. I didn't know the word veterinarian yet, so I claimed that I wanted to become "uncle Christmartin." What I wanted back then in the 1950s was considered extraordinary and no one supported my dream. I come from a traditional Swiss family with a paternalistic economist/lawyer father and a mother who adored him and was the classic Swiss housewife. Uncle Christmartin supported me in my unusual dreams.

There comes a point in Swiss schools where you, and your parents, discuss whether you should go to Gymnasium, which is the academic path for children who want to go to university. The teacher told my parents that normally he did not recommend that girls go to gymnasium, or endorse girls studying there, but he thought that I should go. At this point the discrimination was so clear that it was ridiculous. It meant that I started fighting for my rights from the beginning. Currently young girls assume that everything is okay, and all is equal, and when it begins to get difficult they do not have the resources to combat an unexpected situation.

When I was about 17 or 18 years old I became scared that I would not pass my exams, particularly physics and mathematics, so I went to a psychologist to take a series of tests to check my aptitude. These showed that I was not a typical girl. I scored zero on femininity and had a high score on the potential for math, but a lousy performance nevertheless! The tests recommended that I become a lawyer because of my logical thinking. My father had studied economy and also law and the last thing I wanted to be was a lawyer—I had a very difficult relationship with my father.

Given that I wanted to be a vet, I had already visited the Experimental Research Institute of the ASIF/AO in Davos, by the time I was 15. I spent 2 weeks there, observed the experiments and since then was "nailed" into research science. I knew then that I wanted to be a small animal vet, a trauma surgeon, and work in the United States. I come from a well-to-do family, which meant I was able to do anything I wanted as long as I appeared to be learning. For example, when I was 17 years old I went to the Royal Veterinarian College, London and learned English while doing a placement there. My father told me if I could organize it all myself he would pay for the costs.

My regular visits to the institute in Davos, when I was still at school, meant that I often heard professors and visitors speak at the Davos congresses. I was once able to have dinner, along with another student, with a United States professor who was one of the speakers and had spent some time in Zurich at our faculty. I knew what I wanted and, when he asked what he could do to help, I said that I would like to spend 6 months in his lab working on horse research and then get an internship in small animals. I was able to go to the United States soon afterwards, and achieve my wishes, certainly because of my concrete goals, and focus on the profession, which convinced my future mentor.

An interest in psychology and analysis started early with me. By the time I was 15, I had seen that there were problems in my family structure that I was determined not to repeat. When I was 21, I read Carl Roger's book on encounter groups. This influenced my decision to see a psychologist. He had written about how he'd also seen a psychologist himself, in order to understand issues with his wife. It impressed me that a renowned psychologist would consult with other psychologists on personal issues. It convinced me of the value of embarking on a journey of self-understanding. So I decided to enter Jungian analysis in order not to perpetuate the family history.

Where did you choose to study?

I studied veterinary medicine for 5 years at the University of Zürich and then spent 2 years in the United States at the University of Philadelphia doing first a research fellowship in the equine hospital and then an internship in small animals. This was followed by 11 years working at a private animal clinic in Germany, where I advanced my specialization in orthopedics and trauma medicine. After 12 years I went through a divorce and I decided to change my life. I returned to Switzerland in my late thirties and set off on the path that lead to my current professorship, and my position as dean of the Swiss Veterinary School, University of Zürich.

How did you achieve this after such a long break from academic life?

My first job was as a coordinator for continuing education in the VetSuisse Faculty (50%), while at the same time I began to establish my own research lab. I financed my lab always through soft money, like collaborations with industry, grants from foundations and later competitive grants. I had research questions related to cartilage and bone that I wanted to address. Initially I was unsure about how to tackle my questions and attended a Gordon's Conference, where I spoke with a number of professors about my research plan. Prof. Robin Poole suggested a brilliant method, which I finally used to test my research propositions.

I received my Habilitation in 1999. This was a consequence of my accumulated research and necessary for the next steps in academia. Meanwhile I had founded my own research group called the Musculoskeletal Research Unit (MSRU). Then I became involved in the idea to create a Swiss Competence Centre for Translational Research together with my two colleagues Michael Hottiger from the Vetsuisse and Simon P. Hoerstrup from the Medical Faculty. It is now called the Centre for Applied Biotechnology and Molecular Medicine and it brings together research in regenerative and translational medicine. Working with my two colleagues we made this Centre happen; part of the funding request involved me asking for a professorship. I was initially offered a titular professor position because of my background, but I argued that this was not worth anything and that, if I were given a tenured position, the professorship would be secured for the future in the university after I retired. This was a moment when I did not show the typical women's modesty, but just believed in myself and my abilities. I argued with authenticity for a new concept, and my position was made into a tenured extraordinary professorship.

In recent years you became the dean of the Faculty of Veterinary Medicine? How did that happen?

I have always participated very actively in the faculty. I was actually approached first to become the pro-dean for buildings and resources, and I immediately said yes. Within the Vet School there is an overall policy document called the "Mehrjahres Plan," which contains strategy to develop the faculty every year. I inherited this plan from the previous male pro-dean of resources and buildings. My first action was to rewrite this document by organizing a faculty-wide consultation for my new version. My key purpose was to include the wider groups' needs into the plan, by asking what the rest of the faculty wanted. This approach was such a new concept, that my predecessor told me very seriously that my working methods were wrong; my response was "you've had your time and now I will do this my way." I added changes to the documents, left them visible in the text, so there was full transparency in the process for everybody. It is still done this way now—and ever since that change, the faculty had no arguments about it anymore, which were common before.

In the next step, I was asked if I would stand for election as the dean, to which I agreed. I made an agreement with the other professor, who also stood for election, that if he won that I would remain as pro-dean and, if I won, that he would become pro-dean for Education. We laughed at the prospect that it stayed within the family, since we both have Labradors from the same breeder. In the end, I won with two thirds of the vote. The other third knew that I would not be giving individual people special privileges and would work transparently. Sadly, some patriarchal women worked against me in the background. I have now been dean for 4 years having stood successfully for reelection after 2 years.

What general advice would you give from the wide experience in your career so far?

The Jungian analysis was very significant in my career as it taught me how to understand myself, also to express the real truth about myself and use this knowledge to lead my group.

In addition, my work with animals was very significant to my development because animals will not work with you if you have an unresolved psychological problem. They sense the state that you are in, and you have to develop interactive skills to work successfully with animals. Working with horses is an excellent method to develop mental strength, to understand the problems of operating with too much force and/or too violent an attitude. Many of my PhD students have benefited from this.

My father used to say: success cannot be avoided if you love the things you do, if you are engaged with the topic, passionate, clever, communicate well, work hard, are a fast thinker and self-reflective. These are certainly factors that helped me succeed.

What about your views of leadership?

I have been very impressed by Joseph Jaworski's idea of leadership where you look out for the interests of the group and the community, rather than just for yourself.[2] I take my time to take decisions, look at all options, stay flexible and also wait until the solution comes or is apparent. In my view "don't fight with the sword, use the rapier" and fencing is more like a dance than a battle. The older I get I realize that it is important to fight *for* things rather than to battle the system.

Other pivotal moments?

At a conference a young female surgeon once surprisingly said, "You have survived the system without damage. How did you do this?" My answer was that working with animals shows you quickly that they are not impressed by title and status. They helped me to know myself and what I really can achieve on my own. I also realized early that I did not want to be frustrated as hell by academia and the system. So, I asked myself the question "What do I need to change in myself to cope with the system and maybe change it?"

My experience showed me that some men block ideas from women naturally, so I learned to come from the side using timing, waiting for the right moment and adopting alternative concepts. If I have to fight, I pick my fights and then take it all the way using all of my power. One good example occurred when I first joined the Vet faculty. I had been promised a research lab with my appointment, but the pro-dean—my predecessor responsible for infrastructure and buildings—decided, against the promises of the faculty, that I could not have the lab. After some useless discussions I finally, in a public situation in front of many people, demanded that he provide me with a document that said I would get my lab as promised by the faculty. I insisted that I should receive this document within 48 hours, otherwise I would take the case to the university officials by 12 noon following the deadline, using the grounds that I was being inhibited in my research. The pro-dean responded by inviting me to dinner to discuss this. I remained pleasant, accepted the dinner invitation gracefully, but insisted that nevertheless I still wanted the document within the publicly stated time limit.

He then went to my ex-boss, who was retired, to ask him to persuade me to stop my demands for this letter knowing that we were good friends. My ex-boss's response was that he had no influence on me on this topic and—knowing me—I would have certainly checked out the legal situation beforehand, and that I had been pushed to the limit. I received my letter within the requested time, and as a result, my lab facilities.

[2] Josef Jaworski, *Synchronicity*, The Inner Path of Leadership, 1996 and 2011 (Second Edition).

What other points would you like to add from your experience?

I think we should look at women and science issues differently from the current situation, not simply from a gender perspective and how women have to adapt to the patriarchal system. When people today hear talks that are only on gender discrimination they simply switch off. Young women have to find out authentically who they are as human beings and not just copy male role models. They need to ask questions like: what if my boss does not play along with my expectations? What do I do then to preserve my integrity? We need to empower young women and help them find sisters and support.

In my view the women that operate in a classically male role hurt themselves. If they keep acting like men unfortunately, people call them "monster women," like the stepmother in Snow White who is looking to be the most beautiful of them all. If they keep looking at themselves through the eyes of men, they are always in competition and this is not helpful and makes them very lonesome. We need to assist the younger women in strengthening and empowering themselves to live their lives.

What are the main lessons you have learned over your career?

- Nothing is urgent unless it is a life and death issue—being a veterinarian and mainly a surgeon gives this perspective.
- I forbid myself to worry about small issues, but I am not afraid to acknowledge when things get to me and allow myself to take some days to get over them, until I go on.
- You cannot change what happened to you, but you can change how you work through it.
- I learned to turn a dirt ball into something golden.
- I don't act out of revenge or bear a grudge, I just move forward and leave it behind me.
- I have learned to say no.
- I ask for incubation time on some issues before giving an answer—2 or 3 days.
- You should always empower others and take account of the emotional parts within them.

I also want to add that I still love my job like the first day I began—now I am 64 years old!

Prof. Katharina von Salis, Retired (Swiss)

Retired Titular Professor, ETH Zürich, 1989–2001

Katherina von Salis

Biography

Katherina von Salis was born in Zürich in 1940. She studied geology at the University of Bern and completed a PhD on the geology and sedimentology of the molasse. From 1965 to 1967 she carried out postdoctoral research at the University of Copenhagen, and then at Musée d'Histoire Naturel in Paris on calcareous nannofossils. She was appointed lecturer at the University of Copenhagen from 1968 to 1974, the last 2 years as Head of the Institute of Historic Geology. In 1974 she joined ETH in Zürich and 1992 was appointed as a titular professor in its Earth Sciences Department. She was active in, and presided on, international geological correlation programs and organized international conferences both in earth sciences and gender equality in higher education. She was a member of the Federal Commission on Geology and that of "Berufsbildung"/Vocational Education. In another role she initiated the establishment of the Equal Opportunities Office at ETH and committed herself to working on gender equality in science. From 2001 she worked extensively on gender equality for Swiss Cantonal Universities and for the Swiss Universities of Applied Sciences.

Research area

Katherina von Salis made outstanding contributions to the geology of East Greenland and to calcareous nannofossil and silicoflagellate taxonomy, biostratigraphy, and palaeobiology.

Honors and awards

2012 Anerkennungspreis of the Canton of Grison
2008 Brady Medal of the Micropalaeontology Society
2007 Ida Sommazzi Foundation Prize for work in Geology and Equal Opportunity
2003 Honorary Membership of International Nannoplankton Association
1998 Member of the Royal Academy of Science Denmark and Steno Medal of Geological Society of Denmark

Conversations with Katherina von Salis, October 2015 and September 2018

You have been retired for some years now. How would you like to talk about your career?

As a geologist I like to work with timelines, meaning that I start from the earlier experiences first; though I am also aware that events and incidents appear, and are remembered differently, at later stages of your life. If you use a timeline it places you well in your own history.

What did you study at school that lead you to becoming a geologist?

I went to the gymnasium and then took the specialist path in natural sciences, math and physics, which leads to a Matura, which is necessary to go to university. For some time, I was the only girl in a class of 29 boys, later we were two. At that point only 5% of young people went to study at university, so even that made the next steps unusual.

How did you choose geology as your university topic?

Actually, I was a mountaineer, which made me ask questions about how the Alps had been formed. I had also observed the changing geological strata in the mountains and found marine fossils high above sea level. And then geology was part of the curriculum at high school, too. My choice came from the combination of fascination about the topic and the fact that I liked to be outdoors. I do some walking and e-biking even now, around Silvaplana, in mountainous Graubünden where I live.

At university I was the only women in our year and, back then, we still had fundamental structural problems, for example there were no toilets for women in our building. At this pioneering period in the early 1960s women had to change basic infrastructures to participate. For example, we were not allowed to work on North Sea oil platforms, because there were no showers or toilets for women. I suggested they pay to put locks on the doors, but without success. Another woman geologist, whose grandfather had crossed Greenland with a dog sled wanted to join the ice core drilling in the center of Greenland. This was denied because the United States Airforce at that time was not allowed to transport women. In addition, you had to deal with superstitions that women brought bad luck to expeditions or in mines. I had to dress as a man to get into Austrian mines on an excursion.

Was it considered unusual for you to choose to study geology at the University of Bern?

Absolutely! But I come from a family of unusual women. In the early 20th century my grandmother, Hanni Bay, studied art at private art schools in Munich and Paris. At that time women were not allowed to study at the well-known Art Academies in those cities. She also was an early mountaineer and became a well-known Swiss artist. After her divorce she brought up her three daughters on her own, earning her livelihood/upkeep by providing illustrations for newspapers like Neue Zürcher Zeitung (NZZ) and various magazines and by painting the portraits of well-to-do Zürich people. She was actually a challenging person, but I got on very well with her. I traveled with her on painting tours to Italy, Tunisia and Egypt, which gave me the chance to paint as well. My mother, Charlotte von Salis-Bay, was a well-known Swiss journalist who traveled extensively for her assignments. One of her sisters became an eye doctor. When people in your family are already unusual, or even considered eccentric, it is easier to be different, or to be yourself, within the family.

Where did you study for your PhD?

I did both my basic studies and my PhD research at the University of Bern, meaning that I was there for 6 years. At the same time, I became very involved in orienteering and cross-country skiing and was often underway in the mountains.

In fact, you were a Swiss Orienteering and Cross-country Ski Champion?

Yes, I won the Swiss national championships from 1961 to 1965 in cross-country skiing and once in Orienteering. When I wanted to join the Bernese Academic Alpenclub—which was cofounded by my grandfather, who was also its first president—I was, as a woman, denied membership. Indeed, I also tried to join the rowing club, but there were no changing rooms or toilets for women, so no rowing for me then. I took it up later without such problems in Denmark. It was a time when women were blocked from a lot of opportunities. Some things I could somehow get involved with anyway, others not. You cannot swim against the tide or current all the time. I had learned this lesson when swimming every summer in the river Aare that runs through Bern.

Your next step was a postdoc position in Denmark. Why did you choose to move there?

I met my Danish husband at an international orienteering event in Bern and after my PhD we got married and I moved with him in Copenhagen. At the University of Copenhagen, I started research on calcareous nannofossils, then a new and fast developing field. After a further year's postdoc position in Paris, I was appointed lecturer at the University of Copenhagen in 1968. Here I could finally take part in four geological summer-expeditions to East Greenland. An earlier attempt, as a student from Switzerland, had had no success—they did not take along women. It was also during that time that I participated in three legs of the Deep Sea Drilling

Project (DSDP). These expeditions lasted between 6 weeks and 2.5 months and were spent on a big drillship with over 60 male sailors, roughnecks and scientists. On the first expedition there were also three United States women and they later bought me a subscription to "MS Magazine," the feminist journal founded in 1972, by Gloria Steinem. This introduced me to American feminism. On my last expedition, I codirected as cochief scientist with a male colleague. This happened even though, at 34 years, we were considerably younger than most of the rest of the scientists. We were responsible for the scientific program, scientific crew and how the research was conducted.

Then followed a time when you and your family moved to different places in Europe?

I left Copenhagen for Zürich in 1974, when my husband moved from Copenhagen to Vienna. I took up an appointment at ETH as a senior scientist. Over the next 4 years I had three daughters and commuted between the two cities. My chemical engineer husband worked as a senior executive for Shell and was soon transferred again, first to London and then to The Hague, just after our third daughter was born. After the birth of our second daughter the children lived with him, and we had a British nanny to look after them. Following another stay in Vienna, and later London, where I joined the family, I got a part-time consulting job with Shell. This was not because they wanted me badly, but because my husband had said no to a job in London twice, and they wanted him badly. I had twice tried to get a job at Shell but they, too, did not hire women at the time.

We all moved to Zürich in 1989, where the children did not have big problems to get used to the Swiss school system. I had applied as a full professor at ETH in 1984, but was not appointed. However, I was eventually promoted to titular professor in the Department of Earth sciences at ETH Zürich. I had managed to maintain a considerable research output and also had visiting appointments at the Free University of Amsterdam, the University of Vienna and those of Tübingen and Fribourg. My skills in calcareous nannofossil biostratigraphy also lead to teaching assignments at the University of Caracas in Venezuela and the Brasilian Oil company Petrobras. Working only part-time at ETH, there also was time for consulting work for various oil companies operating in the North Sea and in Asian seas.

It was in 1991 you began to work more formally on promoting equal opportunities for women in science, wasn't it?

Yes, there was a "window of opportunity" to act, after the last women in Switzerland received the right to vote in November 1990. After 1993 the federal minister responsible for the ETH was a woman and she was supportive of gender equality activities. Also, as a titular professor, I was frustrated that I was not able to move my science forward as I wanted, because I was always under the direction of a full professor. It made me focus more on my work in the Equal Opportunity office (25% position).

The then ETH President Jakob Nüesch (1990–97) was ideal to work with on this topic. He made a declaration in front of all ETH professors that he wanted to increase the number of female professors by the time he left office to 30—he nearly reached his goal. The first female professor at ETH had been appointed in 1985, meaning this was a time of very slow change on the appointment of women.

I really enjoyed being part of the equal opportunities pioneering phase at ETH Zürich. Early on it consisted of creating campaigns to change attitudes and using humor to communicate the need to think about gender equality. It was also important to have tactics for handling meetings and achieving policy change. I would work specifically with men on the commissions who were in favor of new initiatives and asked them to make the proposals to the meetings, with them explaining the joint basis for the idea. I learned that I was able to bring about change more quickly this way. Men were more likely to listen to and agree with ideas and arguments from other men than when a woman made suggestions. I find this has changed a bit, but certainly not up here where I live. In 2001 I retired from ETH Zürich. But I was only semi-retired, since I moved into working on equal opportunities in the wider Swiss University and University of Applied Sciences landscape.

What did this involve?

When the new Universities of Applied Sciences were conceived, there was a very strong tendency that only the technical and economic fields would be represented—both were dominated by men. The fields where women are in the majority, namely primary school teaching, nursing, and related professions, were not going to be included. This would have meant that a lot of money would go solely into further education of young men. After all kinds of lobbying with female members of parliament by women from these fields whom I joined, the same opportunities were created to the nursing and education colleges.

In this period, I also had the opportunity to work very constructively with the Swiss State Secretary of Education and Research, Charles Kleiber. I met him at an event in the British Embassy in Bern and after some light discussion asked "So who irons the shirts in your house?" he replied that everyone did it for themselves, because he had daughters and his wife worked outside the home. I knew that we would work well together and the result was the creation of a financial support program for equal opportunity policies for all Cantonal universities in Switzerland. It means that all universities in Switzerland have equal opportunity offices and the money to invest in the topic. There was also funding for national programs.

Has your life experience matched the goals you made for your life?

I never actually had a goal to be a professor, I was just happy doing research. However, my boss in Denmark was a female professor with two children and I observed the way she worked and made things happen, which opened my mind to becoming a professor. If a young woman researcher speaks about her love of science and that she is concerned that becoming a professor means that she will have to leave the lab and science, it is important to realize that there is a real excitement in taking charge of your group and supervising the creation of science this way, and also enabling young people to develop.

I think that often young women in Switzerland were not being taught to "think big"! I once spoke with a young woman who was studying geography and she was quite exceptional in her commitment to the topic and in her insight. When I asked her what she wanted to be, her answer was "teacher" and I said, "Why not be a professor?"

What do you consider were the best moments or experiences in your career?

There were many good moments. Taking part in geological expeditions in East Greenland was hard work but also enjoyable. Developing a field, cofounding and presiding an international specialist-association that promoted young scientists and where cooperation and competition were possible, was an exciting experience. Finally getting into promoting gender equality in higher education in its pioneering phase in Switzerland and Europe was an unplanned, but fine, conclusion to my career.

And finally, what would be your central advice for an aspiring academic?

Go for it. Choose the right partner, have children when they come. Work hard and enjoy all aspects of life.

Prof. Sabine Werner (German)

http://www.mhs.biol.ethz.ch/research/werner.html
Molecular Health Sciences, Department of Biology, ETH Zürich

Sabine Werner, "ETH Zurich/Giulia Marthaler"

Biography

Since 1999 Sabine Werner has been professor of cell biology at the ETH Zurich. She studied biochemistry at the Universities of Tubingen and Munich and obtained her PhD from the University of Munich after having performed her doctoral thesis at the Max Planck Institute of Biochemistry, Martinsried (1989) in the department of Prof. Peter Hans Hofschneider. From 1990 to 1992 she worked as a postdoc in the lab of Prof. Lewis T. Williams, University of California, San Francisco, United States. From 1993 to 1999 she was a group leader at the Max Planck Institute of Biochemistry, Martinsried. In 1996 she became Hermann-and-Lilly Schilling Professor of Medical Research, Martinsried, and from 1995 to 1999 she was also associate professor of biochemistry at the Ludwig-Maximilians-University of Munich.

Research area

Sabine Werner's laboratory studies the molecular and cellular mechanisms underlying the response of the skin and the liver to different types of injury and the parallels between tissue repair and cancer. The lab uses state-of-the-art approaches, including functional genomics and proteomics, 2D and 3D primary cell culture systems, and genetically modified mice for their work. In collaboration with clinical partners, they determine the importance of their findings for the human situation with the ultimate goal to transfer the research results into clinical practice.

Selected honors and awards

2017	ETH Zurich Spark (invention) Award and Ernst Klenk Lecture Cologne, Germany
2014	"Golden Owl Award", for best teaching in Biology, ETH Zurich
2012	Charles Lapière Memorial Lecture, Athens and Elected as EMBO Member
2011	Elected Member of the Leopoldina (German Academy of Sciences)
2010	Chair, Gordon Conference on Fibroblast Growth Factors
2009	CE.R.I.E.S. Research Award for Achievements in Dermatological Research and René Touraine Lecture, Budapest
2008	Cloëtta Award
2001	Chair, Gordon Conference on Tissue Repair and Regeneration
1998	Pfizer Academic Award
1990	Otto Hahn Medal of the Max Planck Society
1987	Kékulé-Fellowship of the "Verband der Chemischen Industrie"

More than 350 invited presentations

Conversation with Sabine Werner, July 22, 2015

Excitement for science and persistence are essential for a scientific career

When did your interest in science begin?

I was excited about science really early. My older sister became a medical doctor and watching her experience made me realize that I did not want to work with patients, but I was interested in the science behind the medicine. At the age of 13 years I knew that I wanted to study biochemistry and work in research. No one influenced me; indeed, the school was sometimes discouraging because some teachers did not consider this as a great route for girls.

I found some subjects in school/gymnasium a bit boring, but I had an enormous curiosity and was most interested in chemistry. I also played in the second German Volleyball league alongside school and my Abitur specialized in chemistry and mathematics.

What was your experience at university?

At that time there were only three universities where you could study biochemistry. I went to Tübingen since I heard positive things about their curriculum. The diploma took 6 years and during that time I spent a year working at the Max Planck Institute for Biochemistry in Martinsried near Munich. I was able to work in six different departments during that year. Although all projects were very interesting, I got particularly excited about the work in the department of Prof. Peter Hans Hofschneider, who studied the mechanisms underlying human diseases, in particular viral disease and cancer. Therefore, I returned to this department as a diploma student after having finished the theoretical part of my biochemistry studies in Tübingen.

As a diploma student I worked on viruses that cause myocarditis. I subsequently stayed in the same department, but chose a project on Kaposi's sarcoma, a tumor that preferentially occurs in AIDS patients. It was the time when the AIDS epidemic was exploding and the patients gave us their biopsies for research. My supervisor, Prof. Peter Hans Hofschneider, was a medical doctor by training and at the same time a famous basic scientist. He set up multidisciplinary work between researchers and clinicians and was a pioneer in moving science in this direction. It was really interesting to work in this group and under the supervision of such an excellent scientist and mentor.

My PhD period was, initially, really challenging. My first topic had been transferred to me from another scientist, but the initial hypothesis on which my project was based, turned out to be wrong. As a result, I had to start from scratch and find a topic myself. It was a real torture at the time and I was very unsure whether it would work out, and I had no idea if I would get the results I needed for my career. It was a difficult and challenging time, but I also learned a lot of wonderful lessons. In the end, my PhD took less than 3 years and resulted in many publications. After completion of my doctorate I stayed in the same group for a few months to finish the project and a final publication.

A year and a half before I finished the PhD I had already begun to plan the next steps. It was always my dream to be a professor and to have my own research group. It was a clear goal for me and I think this really helped because I was never distracted by other options. My main worry was that it would not work out. I knew male colleagues who were absolutely sure that they would become a professor and I think it can be harder for women because they are often more self-questioning.

Anyway, mid-way through my PhD I thought about going to the United States to study growth factor signaling. I wrote to several famous United States labs who worked in this area, but I also considered staying in Europe. However, in the research community in those days, it was a "must," or at least extremely helpful, to have research experience in the United States if you wanted a career in academia. I visited five United States laboratories during my PhD time. I finally chose to work with Prof. Lewis (Rusty) Williams, professor of medicine, University of California, San Francisco. I knew immediately when I visited that this would be the best place for me to perform my postdoctoral studies, since I was impressed by the outstanding science and also by the people in the lab. I also liked him as a person straight away. My aim was to study the mechanisms of growth factor signaling and to learn new technologies and approaches.

During my visit to the Williams lab I was offered a job, but we both agreed that I should apply for a fellowship to fund my postdoc position, because this would be good for my CV and also for the lab's budget. Fortunately, I obtained a Max Planck award (Otto-Hahn medal), which funded my postdoctoral period.

You moved to California from Munich. How would you describe your 2.5 years in the Williams lab?

It was a really tough period since the expectations were high and I worked in a very competitive field. Therefore, I spent many weekends in the lab and worked long hours. The group was very international and included people who all wanted to have academic careers. Therefore, we worked very hard, but we also had fun together. Most of my best friends come from this period of my life. I'd say that for the first 6 months I was homesick and then I had a great time. Fortunately, my postdoc period was very successful. The boss chose the right people to work together. It was not just about recruiting based on a CV, but also about finding those people who fit together. We were competitive, but also supportive.

Before I finished the United States postdoc period I applied for my first research grant. I obtained the grant and could appoint the first student by the time I returned to Germany. I was asked to apply for faculty positions in the United States, but I loved living in Munich and wanted to return to Europe. Fortunately, I had the opportunity to return to Martinsried as an independent group leader. This was a very positive time—I was young and well-funded and it was very enjoyable to be a group leader and to direct my own research projects.

How would you describe your experience with mentors in this period?

I had really good mentors in my career, my postdoc advisor, Lewis Williams in San Francisco, and my PhD supervisor Peter Hans Hofschneider, the director of the Max Planck Institute in Martinsried. He guaranteed me a position in his department on my return from the United States and, when I returned, he allowed me to be completely independent as a junior group leader and he was very generous to me. He guided me through the academic jungle, suggested funding opportunities and put me in contact with important people. During the time in Martinsried I also had the great opportunity to work with Prof. Ernst-Ludwig Winnacker at the Gene Center in Martinsried, who became my third important mentor.

How did the next steps of your career unfold?

I became a group leader in Munich with a great team of five to eight people. It was a wonderful period. I had plenty of time to work with my group, to build up the lab and the research. The results from this time period and the work of this team established my career.

Then another opportunity arose when I was offered the opportunity to replace an associate professor at the University of Munich for a few years. This was my chance to get into the university system and I taught Biochemistry. At the same time, I obtained a Hermann-and-Lilly Schilling Professorship of Medical Research and a foundation paid my salary. At this point my lab remained at the Max Planck Institute and I taught at the university so I was based at both places.

The next challenge, however, was that I could only stay at the Max Planck Institute for 5 years and I needed to find another position. In addition, Peter Hans Hofschneider retired at this time and his lab was closed down. I applied for eight different professorships, and when I was placed second on the candidate list for my first application as a full professor, I began to feel that a full professorship might be possible.

Shortly afterwards I was offered an associate (and tenured) professorship at a German University, but after I weighed up the benefits of staying at Max Planck with my wonderful group, versus this new offer, I decided it was not yet the time to leave. This was a big risk, but I really wanted to go for a full professorship immediately. This was an interesting experience, since many people believed that I would not leave Munich. In various interviews I was asked "Will you be able to run a department as a young woman?" I then received an offer of a full professorship in Germany. I had applied to ETH Zürich, at the same time, but the decision in Zurich took too long and I accepted the first offer and signed the contract. Two days later I received an offer from ETH Zürich.

This was a very tough time and I reflected deeply about what to do next. I thought it was not possible to withdraw from the signed contract, but the ETH Zürich offer was very tempting, with great student/postdoc recruitment options, excellent funding, and an outstanding scientific environment. Colleagues and friends encouraged me to visit ETH Zürich anyway and during the negotiations it became clear that the ETH offer was the offer of a lifetime.

I had to explain the situation to the other university. It had been a very nice offer at an excellent place and they had spent a lot of time and effort on recruiting me. It was hard to withdraw at that stage.

Now your career at ETH Zürich began

It was 1999, I was young and the only female professor in the Department of Biology. Therefore, I was invited to become a member of many commissions and I accepted all requests during the first years, because I was so grateful for being a professor at ETH. However, I also needed time to set up my lab and to build up my group. After a few years I realized that my scientific output did not meet my expectations, because I was so distracted by all the other duties. Therefore, I slowly learned to say "no" to some requests and committee invitations. Although saying "no" is still not my strength, I managed to spend more time in the lab and to reach the scientific top level that is expected at ETH.

Can you tell me something about your research field?

In Munich my group was known for wound healing research with a focus on growth factors in this process. After moving to ETH, I was able to take a risk and expand my research field. We now study the parallels between wound healing and cancer and the mechanisms of regeneration and repair in response to various types of insults. We mainly work on the skin, but we also study liver regeneration. This is an exciting area since the liver is the only organ that can fully regenerate even after severe injury.

At ETH it is possible to start something new without having published in the field using the generous core funding. You can take risks and reap the benefits. I now have third party funding for all of these new projects and I get invited to the major meetings in these areas.

How would you describe your responsibilities as a professor at ETH?

My experience has been really good and positive. There is a lot of responsibility, because you are responsible for the careers of many excellent people. My main concern is to get enough grant money to continue funding all my coworkers.

I usually have 1–2 independent group leaders (Oberassistent) in my lab and this is a conscious management choice because it benefits their careers. Unfortunately, it is difficult for these group leaders to obtain third party funding, because their independence is not recognized. Indeed, a senior scientist, or Oberassistent, may fully work for the professor, but they can also be an independent project leader, it depends on the professor. For me it is important to help advance these young people in their careers and to let them develop their independent research program.

My responsibilities have increased further over the years. I was on the SNSF Research council for 8 years and this took 25% of my time. It was an amazing experience, and I learned a lot about the wider picture of the Swiss research landscape. I am also on a lot of ETH committees, which take time away from research, but these tasks are also very important and interesting.

What excites me most is working on our research projects together with my great team rather than increasing my own personal influence. The biggest pressure at ETH is to keep up the scientific output, and you cannot put your feet on the table. This is not ETH's culture. It is a wonderful culture and we have exciting interdisciplinary collaborations for instance with chemists and engineers and it is great to work with scientists from other departments and disciplines. We had a recent professors' retreat, which lead to the initiation of new collaborations after the two-day meeting and was really great.

Any thoughts on the academic career?

Assistant professors at ETH and elsewhere should have mentors, who guide them through the jungle of the academic system. There also needs to be flexibility in the system to deal with the retirement of professors. The productivity is essential and not the age. Sometimes it could make sense to bring along an assistant professor in the last few years of the retiree's life who would then be a replacement for this professor.

Looking back at your career what would you identify as your challenges?

The most difficult time was at the beginning of my PhD when I had to change topic. Later, it was a big challenge to obtain a tenured faculty position after the 5-year period as a junior group leader and to make the right job decision. A challenge nowadays is that impact factors and metrics in general have become too important, sometimes more than the quality of the work. I chose science because of curiosity and that is what keeps me going.

One of the current challenges for a young scientist is to find a way to be a first author on a paper, because these publications matter for the career. You really need to choose the right person to work with scientifically and also a fair person, who runs a lab in such a way that you can make an academic career.

Another challenge for scientists now is that the first submissions to journals are often judged by professional editors, who receive a large number of submissions of which many are not in their direct area of expertise. Therefore, the decision if a manuscript is sent out for external review often depends on the "taste" of the editor, the chance that a manuscript can get a lot of citations and the influence of the senior author. This is often frustrating, in particular for young investigators. You have to try, try and try again, and our students and postdocs have to learn to handle the disappointment after an editorial rejection. Like many others, I think that journal editors and reviewers need to focus more on the overall scientific quality.

A key factor for me was that I always had a clear career goal—to become a professor in academia. Other people were unsure and thought about other options. I was of course never sure that I would make it, but I am very happy that it happened.

A final point: mentoring of women by women cannot always be done and the few female professors get overloaded with mentoring tasks. All my mentors were men and they were all great. I mentor the people who I accept into my lab, whether they are men or women. I still mentor many who have left my group, even those from a long time ago and I continue to write letters of recommendation.

Prof. Marcy Zenobi-Wong (American)

http://www.biofabrication.ethz.ch
Tissue Engineering and Biofabrication Laboratory, Institute for Biomechanics, Department of Health Sciences and Technology, ETH Zürich

Marcy Zenobi-Wong, "KellenbergerKaminski"

Biography

Marcy Zenobi-Wong is an associate professor of tissue engineering and biofabrication in the Department of Health Sciences and Technology (D-HEST). She was assistant tenure-track professor in D-HEST from 2012 to 2017. In 1985 she received a bachelor's degree in mechanical engineering from MIT (Massachusetts Institute of Technology), Cambridge, MA, United States. She moved to Stanford University to continue her studies and obtained an MSc (1987) and a PhD (1990) in mechanical engineering. In 1991 she worked as a postdoc at the University of Michigan. She was a group leader at the University of Bern from 1992 to 2002 and habilitated there in 2000.

Research area

Her main research focus is cartilage engineering and regeneration: the success of cell-based therapies for tissue repair is dependent on the ability to reliably control the growth and differentiation of stem cells used in the treatment. The interaction of cells with the extracellular environment can have a potent influence on cell fate. The nature of biomaterials (composition, charge, and stiffness), ligand and growth factor availability, oxygen tension and the presence of mechanical and electrical signals can potently induce the desired morphology, cytoskeletal architecture and biosynthetic activity. The group engineers 3D cellular systems to mimic different extracellular environments and uses photopatterning and bioprinting techniques to construct complex cellular structures. Her research lays the foundation for developing successful clinical strategies for tissue regeneration and repair.

Honors and awards

2008–10 Marie Heim-Vögtlin Award

Conversation with Marcy Zenobi-Wong, May 28, 2015

So how did you become a scientist?

Actually, in another world and born to another family, I probably would not have become a scientist at all. I am the third of seven children and was brought up in the suburbs of Boston where my family was one of only two Chinese American families in town. My parents were quintessential "Tiger" parents: very strict with high expectations of good grades and lots of practicing (piano, skating, etc.). We were expected to excel and there was unspoken rule we were not to disappoint them.

My father was born in New York City's Chinatown, while my mother immigrated to the United States from China to study at college. We were an MIT family. My father and all his brothers went there and my father credited the school with raising him up in society. After my father's PhD, he went on to establish his own successful company, which manufactures semiconductors. He valued MIT and an engineering education above everything. He wanted, or expected, all his children to go to MIT or Wellesley College (a women's liberal arts college outside of Boston). My mother did not work outside the home once she had children.

All seven children have higher degrees, three became medical doctors and three are in engineering and business. My youngest sister studied Art History, but that was an exception.

I had no desire to go to MIT. In fact, I was accepted at other Ivy League schools, but I eventually gave into family pressure and went to MIT. My two older sisters and my youngest sister went to Wellesley. Without the nonverbal pressure and expectations from my parents, I would have studied art or literature. I think, in the end, my route toward science gives me a different perspective.

What then was your experience at MIT?

MIT is a haven for nerds who love engineering and science. I studied there for 4 years, lived on campus and spent a lot of time trying to fit in. I eventually settled on a double major in mechanical engineering and english literature. It was an academically tough environment and I studied hard without really thinking about what I wanted to do.

You left MIT after graduation to do a PhD at Stanford University?

Yes, I was given a research assistantship at Stanford to support my PhD studies. My PhD was in mechanical engineering and was focused on the then new field of mechanobiology. The fact that mechanical loads could influence biology was a radical view then, though today it is an established field.

The West Coast environment was like a foreign country to me, compared to the greater Boston area where I grew up. I met my Swiss husband there, who was also a PhD student, and then we both moved to postdoc positions directly after graduation. I was at the University of Michigan and he was at the University of Pittsburgh. We visited each other every other weekend, driving between the two cities.

Did your postdoc work continue on from your PhD research?

The group in Michigan was different because it was an experimental orthopedic lab. I had already felt that something was missing by working solely on the computer. It was really exciting to be in a wet lab and the theoretical work from my PhD gave me a useful context in which to view the data.

What were the next steps? How did you come to live in Switzerland?

We both looked around for faculty positions after the postdoc period, then my husband got a fellowship offer at EPFL, Lausanne, in Switzerland, which was a springboard to become an assistant professor. The first years in Switzerland were hard for me being so far away from family and friends.

What did you do when you came to Switzerland?

We arrived in Switzerland in 1992. I found a very interesting postdoc position at the University of Bern. The labs there focus on treatments to repair cartilage, which was exactly my field. I was very happy with this position and at that point we lived in Fribourg and commuted in opposite directions each morning. However, as soon as my husband was offered a professorship at ETH Zürich we moved to Zürich.

The first two children were born in 1994 and 1996, when we moved to Zürich. After they were born, I took the standard maternity leave (14 weeks), but never stopped working. I went back to work at 60% but continued to publish because I had postdocs and a technician. My husband and I shared the parenting. I commuted to Bern for 3 long days, and my husband took the children to kindergarten and picked them up. The other 2 days I spent with the children. The childcare options were so minimal back then, that we found our Zürich apartment based on where we could get daycare places for the children.

I have always felt it is a privilege to be able to work and be a mother. It was always good to have this balance for when things got frustrating at work or tiresome at home. At that time, I had a senior scientist position and I taught Histology to medical students, and I could apply for my Habilitation.

When did you make the move to ETH Zürich?

I stayed in Bern until 2003 and then got a position at the Institute for Biomedical Engineering, ETH to establish a master's program. This was a hybrid position and more administrative than my previous job. It was not an easy environment to carry out research and publish and I have a gap in my publication record from 2003 until 2010.

In 2004 and 2006 I had two more children. I successfully applied for a Marie Heim-Vögtlin (MHV) grant, which was established to assist women returning to academic careers. This grant was an essential element in enabling me to continue my science career.

It is not easy to keep up with research output working part-time. My first step was to hire PhD students so results were being produced while I was not always physically in the lab. Then I was successful with an SNSF grant application and subsequent grants. Before I received the MHV award I had to write the grants, to be in the lab myself and there was no possibility to delegate anything.

Your appointment as assistant professor at ETH Zürich was unusual, wasn't it?

Yes, that's right. At ETH Zürich assistant professors are generally appointed under the age of 35. I was very fortunate that the president at the time, Ralph Eichler, made an exception to this rule. In 2011, I was appointed as a tenure-track assistant professor at the age of 48 years. I hope my case will encourage the age rule to be interpreted more flexibly.

Being older and "returning to academia" has some advantages as you've had more time to reflect on what is really important and what are fads. Obviously, the ability to return to a field after a 7-year break depends on your field. For some fields a long break would be a death sentence, but orthopedics is a slower moving field. The outcomes and measures are imprecise and identifying treatments takes time.

I love my job. Science is like a puzzle and I enjoy putting things together to find patterns and to make sense of my findings. I had resigned myself to being a senior scientist for the rest of my career. We decided to stay in Switzerland for my husband's career, which meant my chance of my becoming a professor was small. I feel very privileged that it worked out.

What was the biggest challenge?

The hardest part was to try to do competitive research with minimal resources, no permanent position and no coworkers. Once I became a professor, things became a lot easier since I could create a team with a critical mass.

Marcy Zenobi-Wong, was awarded tenure in May 2017, becoming an associate professor at ETH Zürich.

Outcomes and lessons from the conversations

A professorial career

What has emerged from these conversations is the unique nature of each professor, their disciplines, individual choices, and the routes they took to their current positions. However, at the same time, a number of uniting and interesting characteristics started to appear, which apply to anyone interested in a research career, regardless of gender. It was possible to identify three unifying elements necessary for the career: an unstoppable drive to find your research interests and questions, coupled to a persistent interest in the exploration and development of your field. The second element is the characteristic of individual inner strength and resources, which means an openness to new possibilities, preparedness to try for everything, persistence in the face of challenges (both intellectual and career-wise), and ongoing positivity and resilience. The third element is that these professors grew their careers alongside people and/or environments who supported their goals, objectives, and ambitions, whether it was through being open minded about life choices, encouraging the next steps, or providing necessary funding. In the section that follows these factors will be explored in more detail.

Finding your research field and your questions

Only a few women professors realized early that they wanted to be research academics, or even professors. The majority loved school, and explored naturally topics that interested them, but it was not clear for many years that this would be a career to pursue. Some, in the early years, were actually disinterested in school and what it might offer, and tried out other jobs, or courses, until they found the topic that would occupy by them for the rest of their career. The pioneers, stepping into academia when the cultural climate assumed that further study was mainly for boys, were faced with being considered unusual, perhaps strange, for choosing the scientific routes at high school.

Nicola Spaldin argues that finding your individual topic or questions is central to your research career.[1] This process does not usually appear as a one-moment insight, but rather through organic growth as you try different topics within an academic area, experience different research environments and realize what is

[1] Spaldin, N. (July 3, 2015). *Find your most interesting question.* Working Life, Science. www.sciencemag.org.

Inspiring Conversations with Women Professors. DOI: https://doi.org/10.1016/B978-0-12-812346-1.00002-0
© 2019 Elsevier Inc. All rights reserved.

interesting and what is not for you, but still feel compelled to go on and to find your topic. It might be that others encourage you in directions that just click with your enthusiasm for discovery and exploration. In these conversations, we see that this is happening across all the fields and specialisms: biology, physics, chemistry, mathematics, the psychology of science education and learning, geology, engineering, theoretical chemistry, or cosmochemistry, to list a few. Their journey was to link intellectual curiosity, interest and research skills to a topic that presented a whole career of exploration. Some women professors became theoreticians in their fields, others are experimentalists, and others need the challenge of both working with theory and practice.

All of the professors found the specialism that drove their interest and exploration, and the questions they want to use their talents to address. As Ursula Röthlisberger says, if a topic resonates with your heart and mind, and your research expertise gives you the freedom to continue exploring all aspects of this area, you also have a strong perspective and counteraction, when having to deal with difficulty and challenging times in academia. The more the research expertise grows the stronger the links within research communities and university environments. Finding your question is the first, essential, aspect of becoming a professor, and then you need to explore whether this topic is enough to maintain your interest and motivation to grow a career.

Shana Sturla, found that her topic, after considering a range of specialisms in chemistry, is Toxicology; which combines her interest in chemistry, the environment, and society. She urges young researchers to really invest in their research projects without hesitation about where it will take them because, even if you do not choose an academic career, or it does not choose you, you will have given your all. Developing the skills of full and heart-felt application to a topic, gives you strong experiences that will benefit you for all life's choices and whatever steps you then take to find your own career.

Being suited to the position of a professor

Achieving a professorship requires you to be an independent person, or find ways to develop the necessary independence, so that you can work on your own to build up the early career results. You can seek advice, and the support of mentors, but the key to making progress is finding the drive within yourself for this type of life.

Academia has to also fit with your character, you have to find that pursuing research goals brings happiness, that you want to devote considerable time to the process. Clara Saraceno spoke of the excitement of having research freedom while working as an intern in the United States. "I loved the freedom I had to check out a number of directions with a project," which meant that she was compelled to work evenings and weekends on problem solving. She had not decided that academia was the way forward then, but the joy she felt from the freedom to explore and use the investigative process was something she would recognize in herself later as she decided to try for an academic career, rather than move to industry.

You also need to find happiness from all the aspects of academia: challenges such as a greater teaching load, supervising young people, and growing a group in the next steps in your career. Emanuela del Gado describes how she had not initially expected to enjoy teaching or supervising, but when she experienced working with young students on physics problems and observed them, gradually, taking the responsibility for the problem-solving process and outcomes, and also finding joy in their achievements—she knew then that this aspect of being a professor was something she could not do without. Many professors move from research institutes, or choose not to join them, because they enjoy the experience of working with, and supervising, young people during the research process.

A further characteristic common to all the professors was that they created early and continual opportunities for themselves (internships, placements, opportunities to learn new languages). Many worked for a short time in laboratories in research institutes, in industry, and even in universities, before and during their undergraduate and masters experience. As young women they were testing whether they enjoyed a range of potential working environments and, at the same time, already building practical research experience. They were interested in new life opportunities in general, in traveling and in learning extra languages abroad, even if these experiences were not always related directly to career possibilities.

This quality of preparation in advance meant that many applied, at the earliest stages, for doctoral grants to finance their PhD, and later fellowships for the post-doc period. As undergraduates some of them were already seeking out the potential financial support that exists for young people's careers. They did not wait to apply for research funding, but rather reached out to apply for next step career awards, for advertised jobs—even when getting these grants seemed to be impossible, they still decided to try. Many found that daring to apply brought successes early, but they were also ready to learn from rejections by seeking feedback on what went wrong, and then trying again if the grant applications failed.

A final quality emerging from these conversations is that all these careers have been underpinned by the fact that our professors were looking to find happiness in a wholistic way, taking account of partnerships, their families (if they decided to have one) and also the more general environments they lived in. A career as a professor is a way of life, and in terms of making decisions it was important that the other aspects of life were also balanced and made whole in this career. The decision to take a professorship, or to step on the path to make it happen, was affected by partnerships, culture, natural environment, growing a family, meaning there were also some compromises alongside the drive to make a mark in their research expertise. All of the women professors found the belief to take the steps to make their career aspirations real, and many focused on the positive elements in their environments (academic and personal), that encouraged them and then underpinned the achievement of their goals.

Influences in the surrounding environment—origins and turning points

Again, the conversations show that the women professors have a wide variety of origins, academic experiences, and different life choices. It is tempting to try and find a summarizing view of how the various environments influenced each stage of their careers—for example to find a sentence about the high school period that says that their school experiences were enabling, that they all attended schools with open views on young women's careers, that the teachers were great mentors, and the families were only encouraging and open to all choices.

In reality life is much more complex than such sweeping statements and, in the end, one can say with truth that in high school these young women, as individuals, found their next steps through dealing well with the circumstances in their surrounding environment. The women professors pursuing an education in the mid-1960s were pioneers. Katherina von Salis came from a family of unusual women, which meant she did not question what she wanted to do, despite being an absolute rarity in the academic high school system and studying in a university department where there were not even female toilets. Brigitte von Rechenberg also had to deal with a culture that did not believe that girls should take the academic route to gymnasium, but something about her qualities in school persuaded the influential teacher to recommend to her parents she take that route, and she always knew she wanted to be a vet, which meant that university education was an absolute requirement. Elsbeth Stern had the experience of coming from a farming family where the tradition was to educate only boys, but she was supported by a teacher and aunt who influenced positively her family's view of education for girls. These are examples of significant steps, and turning points taken, that are part of an influential whole. They are underpinned by the clarity and determination of these professors as young women, alongside the necessary support to make their wishes happen.

These same moments, or turning points, can be found in the lives of all the women professors, even as it became more acceptable, and then became normal, that young girls should continue into higher education, and even natural science degrees. The key is that these turning points often arise from personal decisions made by individual professors as young women, and their decisions may include the multiple influences listed: to identify internally an early goal to become a professor (Elsbeth Stern, Sabine Werner); to realize that research on the goals of natural sciences is a fundamental and an authentic life for them (Ursula Röthlisberger, Eleni Chatzi, Nicola Spaldin, Emanuela del Gado, to mention a few); to find that you are so compelled by curiosity and fascination with your topic that you keep taking/finding the next steps in academia (Olga Sorkine-Hornung, Paola Piccotti, Salomé LeibundGut, Isabelle Mansuy, Stefanie Hellweg, Natalie Banerji, Clara Saraceno, Marloes Maathius, Shana Sturla, Effy Vayena, Marcy Zenobi-Wong); to aim early for a career that would give financial independence, but then find a research topic that becomes compelling (Ursula Keller, Rachel Grange). For a number of professors, the turning points, and the way forward, were particularly hard won, after struggling with early choices and jobs that did not, ultimately, bring fulfillment (Ulrike Lohmann, Maria Schönbächler).

For some professors the support of influential teachers, academic mentors, and supervisors, made a huge difference in decision-making, but for others each step was found by themselves alone, all steps involved a lot of self-reflection and deep internal learning. The participants in this book did benefit from opportunities, but many made the opportunities for themselves. The next section explores what the professors said about the role of teachers, mentors, and supervisors.

Influences in the surrounding environment: mentors, supervisors, and colleagues

All of the female professors had positive experiences within their academic communities, whether in the university environment or the wider international community related to their research expertise. Some professors had both. In these conversations they spoke about the people at every stage of their career, who made a difference by their generosity, encouragement, and collegiality. In addition, given that most of these women developed careers in scientific disciplines the majority of these encouraging academics were men.

PhD and postdoc supervisors introduced them to research communities, opened their networks to them, encouraged their ambitions by enabling the development of independent research projects, nominated them for prizes, and suggested they should consider applying for ambitious jobs. Effy Vayena describes how one key supervisor's generous support and influence changed her life and career by opening up an important postdoc position in the field of bioethics. Other supervisors were also influential in giving independent intellectual space for the young researcher to grow, to find their own questions and stand on her own two feet. Certain professors also learned that, if there were not going to receive detailed supervision, they had to become proactive and develop the strength to achieve what they wanted.

The professors in their thirties and early forties have had positive experiences with colleagues and senior leaders who supported them and offered advice through the career steps such as joining a department, establishing a group, preparing for tenure, and seeking new positions. Working in an environment where their colleagues and senior professors created a positive and encouraging atmosphere supported the challenging career period of transforming from assistant professor to a tenured position.

However, not all research communities and departments have been friendly and welcoming. A number of professors had to find ways to establish themselves and their groups in competitive environments, which were not naturally welcoming, where it was easy, as a minority, to feel your voice diminished and silenced. These experiences have needed resilience, strength and the determination not to depart from a uncomfortable situation. They have also been helped when people in leadership positions have recognized, and dealt with, any clear problems constructively. In these circumstances the networking opportunities established by a Women Professors Forum can support women academics going through various career stages such as applying for tenure, working to gain research or departmental funding, and to develop influence within your institute or school. This is where female colleagues from other institutes and departments can bring support and also important perspectives during the transitional phases of a career.

Influences in the surrounding environment: partnerships and children

The central drive of these conversations was to speak about the women professors' careers, but if they chose to discuss whether they had dealt with a dual career situation as their careers progressed; this added depth of perspective about their lives. What emerged from the conversations is an inspiring picture of many partnerships open to the variations and possibilities of life lived, and open to remaining flexible about how to find joint happiness, where to set up home and how to make their careers happen. These partnerships give mutual support and value a career that involves the development of intellectual pursuits. At times their lives proceeded forward with one member of the partnership making compromises, in order to enable steps to be taken in the other's careers. Many professors communicate how their partners were an essential support on their journeys to professorship.

More than half the professors have children, and their partnerships are underpinned by the fundamental belief that parenting is an equal responsibility for both parents. The families have been, and are, supported by using excellent university childcare facilities, by individual childminders or nannies; with both parents ready to step in, in case of illness or unexpected eventualities. Within these families there are no preconceived ideas of the roles each parent should or ought to play. Of course, their experience is set within cultures, which have unspoken expectations about how parenting should happen, which can mean that they live with a range of uncomfortable societal views on the choices that they, as working parents, make.

It has to be added that not all professors in an academic environment have partners, have families, or are in traditional partnerships; and the diversity of those who participate in academia continues to grow.

Specific challenges identified in an academic career

The professors identified what they perceive to be a number of general challenges to aspiring young career researchers at different stages of the career. Their advice focused around the topics: establishing your publication record, dealing with temporary postdoc contracts, and managing career transitions.

Developing a publication record as early as possible

It is clear that it is very important to invest in the intense process of research output that characterizes a PhD period. However, the ability to have publication(s) in peer-reviewed journals is determined not only by one's own individual efforts, but team work, the way a research group is supervised and the current journal review process. Sabine Werner identifies how important it can be that a research group is organized to ensure that young PhD students can achieve first author publications, are named in joint research papers and have the opportunity to present their work to conferences. The ongoing recognition from the lead professor that this process matters, and sometimes has to be controlled and rectified, is vital to the success of younger researchers. The encouragement and recognition that the professors in this book received at this early stage, combined with their ability to produce the research output and write the papers, was vital to career development.

My experience of working with and talking to young academics across universities and research institutes indicates that this process can be complex and problematic at times in academic groups and it needs care from leadership. The competitive environment, if it is not subject to checks and balances, can mean that those with the most power in research groups can take all the accolades and exclude others from recognition and merit.[2] Research groups lead by professors who are aware of these potential dynamics can make a huge difference to a young person's career progression or, indeed, motivation to remain in an academic environment.

Sabine Werner identified a further factor under this category, which arises from the changes to academic publishing in the last decades. Rejection of a paper is always a difficult experience, but previously PhD students received feedback reports identifying both strengths and weaknesses, alongside the rejections. A shift in important journals to professionally appointed first readers, who can trigger a direct rejection, means that PhD students are receiving less valuable feedback for resubmission.

Shana Sturla also identified how the growth of interdisciplinary research, and innovative research directions, can often mean that new fields and the associated findings are only slowly accepted for publication by reviewers. She highlights the challenge that young academics face, when paper rejections come from anonymous older academics, who have little experience or insight into new fields. It is important to support young researchers at this point, by explaining the complex dynamics of the publishing process, and the need for persistence in communicating their findings. This is a challenge of innovatory fields, but such research can have large benefits in the long term. New ideas and directions, despite initial opposition, have influence and shift perception and outlooks in a field. The conversations with Paola Picotti and Marcy Zenobi-Wong give examples of the fruitful outcomes of new directions in research.

The postdoc experience

Many professors worked as postdocs in the transition period when they were building up their research expertise, and following this time took the next steps to research group leader, or professor position. For some professors, it was only a one-year position to write their research papers and consolidate findings. For other professors this was a time when they found their new research directions or questions. For many the postdoc time lasted for a longer period, where the professors grew the expertise in their research field and moved into positions as senior postdocs in a research groups. The length of a postdoc experience can depend on the chosen field but it is is also affected by the scarcity of professorial positions at university, which reduces the opportunities to take the next step.

This postdoc period where early career researchers take on temporary contracts as postdocs has been identified as "precarious" and especially when this period coincides with a decision to become parents.[3] A number of the professors became parents during their postdocs and their experience shows how vital it was to their success that their supervisor, or research group leader, had a flexible approach to postdoc researchers who became mothers, in terms of giving them the independence

[2] Donald, A. (February 2017). *Building a human workplace*. Optics & Photonics News.

[3] See *Harvesting talent: strengthening research careers in Europe*, an investigation by the Leading European Research Universities (LERU) network, 2010. www.leru.org.

to manage their research schedule and family responsibilities. The key factor was that as long as the postdoc mothers delivered their research output and goals they were free to manage their time schedule themselves. This flexibility, and the open-minded approach from their supervisors were a key factor to these women proceeding with their careers, and at the same time becoming parents. Finding a way to juggle motherhood and a research career will be discussed further later in this chapter.

Sabine Werner also argued for a more flexible view of career progression in academia in general. For example, she argues that the university environment should be more open to different paths to a professorship, for example she suggests to make it possible for senior scientist positions to also be a stepping stone to a professorship. Often, within larger research groups, young group leaders are appointed as senior scientists, but there can be a challenge for these young career academics to demonstrate the independence of their research from their professor, which can then hinder their progression. The key to widening the routes to career progression is to change the way these positions are handled by the professorial community.

The postdoc period is actually the time when the majority of young people leave academia, departing from university naturally to undertake other fulfilling careers in industry, government, banks and consultancies, to mention a few. Indeed, universities such as ETH Zurich, also support strongly students who want to become entrepreneurs and establish spin-off companies.[4] There are relatively few professorial positions and for these jobs the academic community seeks to retain a number of the most talented, by expecting them to demonstrate not only research expertise but also mobility between countries and research groups whilst also building up teaching experience. Young people pursue their academic careers in an environment where there is a scarcity of permanent jobs across the world. As more young researchers are now committed to partnerships, where there is dual career situation, they need to find a compromise that enables both to have a fulfilling job, and to live in the same place. This often means that many leave academia during the postdoc period. It can become impossible to live with the career uncertainty, a dual-career challenge and scarcity of permanent positions.

Our professors navigated this challenging period successfully, though for some it was longer than others, and involved times of questioning, self-reflection, discussion, and examining other alternative job directions. Such experience is never easy and often the most important characteristic during this time was their ongoing resilience and persistence in finding ways to keep the research going.

Managing career transitions and the tenure process

During the four years of this book project some of the professors were promoted to a tenured position, even though at the beginning not all had tenure-track professorships. They had taken the risk, and opportunity, to pursue their research careers by accepting temporary contracts, which meant that after 4 - 6 years they would have to move on. However, the result has been that, taking these risks to accept a

[4] For more information see www.ethz.ch/en/industry-and-society/entrepreneurship.html.

temporary contract bore fruit because of their successful research performance which yielded such positive results that they went through the tenure process and were awarded tenure. Many professors repeated the phrase, "You have to dare to try." You have to take the next steps, however risky, because you cannot know what positive possibilities are, potentially, ahead.

It is inspiring to report this success with managing the tenure process, which happened across a range of disciplines. The experience during this assessment process is very stressful and challenging for all those professors going through such a transition. A professor is dependent not only on having excellent results, but on departmental colleagues who will rate their performance, and on international experts who will, anonymously, judge their future. This is where it is vital for young professors to be embedded strongly in their communities, with good relations at both the departmental and university level.[5] The importance of integration into your local community cannot be underestimated, because it is vital to tenure success. Indeed, young tenure-track professors, who are also young parents, can be vulnerable to losing contact with their department at this point, because of the intense investment needed in caring for a young family when the children are small, and the potential loss of connection with events in the community can affect voting choices to support promotion. Formalized networking efforts, and the support of colleagues, are crucial during a tenure process that can take between 1 and 2 years, meaning a phase of uncertainty in terms of research group development and outcomes. At this point, young professors often apply for many other professor positions in their community across the world; it increases their job options, and can help with the negotiation process, which takes place once tenure, and a permanent position, is awarded.

Specific issues for women in the academic environment

It was not possible to write a book, such as this, without asking the professors about their experiences of being a woman in academia, and whether this brought them challenges in their career. Gender statistics internationally demonstrates that the share of women in academia decreases dramatically at each step of the career ladder, with them being a small minority at the professorial level, even when there is a large participation of young women at earlier stages of the career. Box 1 is taken from Advice Paper No. 23, of League of European Research Universities (LERU), which uses the *She Figures* 2015.[6] This Box contains Figure 1, which illustrates the gender distribution in academia in Europe showing that the percentage of female professors, across all disciplines, remains at 21%. In natural science fields, such as Physics, Mathematics, and Computer Science, the share of female

[5] A blog from an engineering professor, Leslie M. Phinney, working at Engineering Science Center, Sandia National Laboratories, in the United States, gives her experience of the tenure process: https://www.insidehighered.com/advice/2009/03/27/what-i-wish-id-known-about-tenure.

[6] Box 1 is adapted from the LERU Report, *Implicit bias in academia: a challenge to the meritocratic principle and to women's careers – and what to do about it*", January 2018, p. 7 and She Figures 2015, p. 127, Figure 6.1. Source: Women in Science database, DG Research and Innovation and Eurostat.

professors stands at a lower level, at just over 10%.[7] Europe-wide there is a significant commitment at government levels, for example the State Secretariat for Education Research and Innovation (SERI) in Switzerland,[8] and in funding bodies such as the European Research Council (ERC), and the Swiss National Science Foundation, and in universities, to uncover the roots of this persistent imbalance in

Box 1 Male and Female Academic Career Progression.

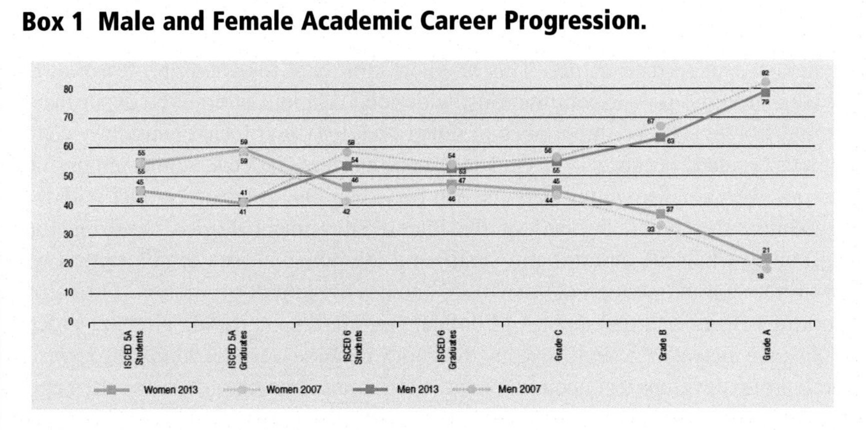

Figure 1 The "leaky pipeline" or "vanish box" negatively affects women in academic careers across Europe.

As Fig. 1[9] shows 59% of graduates but only 21% of (Grade A) professors[10] are female, whereas 41% of graduates and 79% of professors are male.

The graph also shows that progress between 2007 and 2013 has been minimal.[11] The number of women employed as heads of universities or institutions entitled to deliver PhD degrees is 15% (EU-28 average, 2014) showing a similar growth rate to that of women Grade A academics.[12]

[7] ETH Zürich gender statistics confirm the figures for natural sciences see Gender Monitoring, part the website of the Equal Opportunities Office: https://www.ethz.ch/services/en/employment-and-work/working-environment/equal-opportunities/strategie-und-zahlen/gender-monitoring.html.

[8] A recent 2018 report in German, under publications, on the State Secretariat for Education Research and Innovation (SERI) website, https://www.sbfi.admin.ch/sbfi/de/home.html.

[9] Etzkowitz H., & Ranga M. (2011) Gender Dynamics in Science and Technology: from the "Leaky Pipeline" to the "Vanish Box", Brussels Economic Review, 54 (2/3), 131–147.

[10] Grade A "corresponds to the rank of full professor in the majority of the countries, or otherwise represents the highest post at which research is normally conducted (*She Figures* 2015, p. 130)."

[11] Currently (i.e., in 2017), *She Figures* 2015 gives the most up-to-date large-scale data.

[12] *She Figures* 2015 (p. 142, Table 6.4 concerning 2014).

numbers and promotions, and to institute a range of measures to remedy the situation.

When asked if they had experienced problems or disadvantages in their careers from being female, many replied that they did not consider that their gender created obstacles in their careers. For these professors, their focus was on the challenges and excitement of their particular field, making research progress, the specific experiences of encouragement and motivation that lead them on, and the particular people who contributed to their success. The professors with the longest careers described a diverse range of experiences. The pioneering professors, such as Katerina von Salis and Brigitte von Rechenberger had to deal with environments unused to, or even not open to, the participation of women in higher education. Elsbeth Stern spoke of dealing, in the early days of her career, with "old boys' networks" that excluded experienced female academics from professorships and also excluded competent men who were not part of those networks. Other senior professors identified periods when they felt they were working in an unwelcoming environment in their discipline, or faced attitudes in their communities that initially blocked their progression and promotion. However, other senior professors did not speak of such challenges at all. The more recently appointed and promoted professors described their positive transitions to professorship, speaking of how these very intense career phases, proceeded on the basis of both the support of their academic and departmental communities and because of their own significant investment and planning to produce the successful results for their research careers.

Being a female professor

Two specific and current challenges to being a female professor, were raised, in different ways, by a number of people. The first challenge is handling the burden of significant administrative and committee work, which being a female professor has the potential to attract. The second issue arose from the discussion by some professors of their experiences of witnessing attitudes and behavior in decision-making bodies, which they described as implicit and unconscious bias against women. They raised the point that patterns of negative behavior and attitudes exist, which if not questioned, could have a detrimental impact on the recruitment and promotion of female academics.

The first challenge, that women professors attract significant committee work, is a very practical one, and arises from the fact that female professors are in the minority in the academic environment. This means that women professors can be given extra responsibilities because of the need to have female representatives on all of the important committees in universities. These types of duties can grow out of control if not managed well. Professors talked of how the responsibilities bring valuable networking experiences, insight into university structures, and can involve considerable responsibility and power, but by sitting on too many committees these duties can become a significant burden. The amount of work involved can impact the research output of a group if it is not handled well, or a balance is not found. Indeed, the scarcity of female academics can even mean that female professors are invited onto committees and review boards outside their country, adding more administrative workload. Learning to decline, or manage, this invitation process has been vital to many of the professors, Sabine Werner and Natalie Banerji talked about the important process of learning to say no.

Undertaking significant administrative duties on the tenure-track can have a negative impact on success during the tenure process, especially given that the academic community is measuring success via research output and not via "academic citizenship."[13] The professors who experienced a reduced research output, while taking on too many administrative responsibilities were fortunately young newly tenured professors, but from experience they advise strongly against agreeing too easily to these duties. Sometimes, as a young professor, it is difficult to say no, but being warned by colleagues of the negative impact of being overburdened during a transition period, such as tenure review or returning from maternity leave, provides an argument to decline or postpone these responsibilities. A further solution to help ease the administrative burden in general, came from professors who spoke about the importance of working with well-informed male colleagues who will advocate independently and positively for women on these committees. This could be one way to spread the responsibility across the significant decision-making bodies.[14]

The second challenge raised, more often by professors with longer careers, was a concern about the existence of what is called "implicit and unconscious biases," meaning unspoken underlying attitudes and behavior in academia that can potentially hinder the appointment, promotion, and retention of female professors. These professors had participated in university commissions, such as appointment and selection committees. In one example a professor describes sitting on a recruitment panel and hearing quiet women candidates described as colorless or mousy, while confident women were considered to be aggressive or difficult. In contrast, in the same setting, a male candidate was described as "full of hot air," but still made it onto a shortlist of candidates. In this particular case, the male chair of the search committee, responded positively to the female professor who pointed out these discrepancies in assessing the candidates, and it made a difference to the shortlisting process. However, if female professors are left to raise these issues in each situation, little progress will be made, which means that a leader or chair who is aware of the effects of implicit biases is very important, particularly when the progress of women into professorial and leadership positions remains slow.

[13] Macfarlane, B. & Burg, D. (July 2018). *Rewarding and Recognizing Academic Citizenship*. The Leadership Foundation for Higher Education. www.lfhe.ac.uk.

[14] Bagues, M., Lylos-Labini, M., & Zinovyeva, N. (2017). Does the gender composition of scientific committees matter. *American Economic Review*, *107*(4), 1207–1238.

A second example under this category of unconscious behavioral patterns, was the experience of professors witnessing women's voices and views being ignored in committee discussions, only to find that if a male colleague subsequently raises the same point exactly, his view is recognized positively. Such regular patterns of behavior, can be demotivating if you are a minority member in decision-making bodies, and research shows that the negative impact of these experiences can increase over time.[15] While the female professors in the book did not dwell long on situations where they felt disillusioned or discomfited, the potential impact of implicit and unconscious biases in university settings, and in funding agencies, is the focus of recent reports, research, and policy making. The 2018 LERU report in particular, makes suggestions of what can be done to address biases at all levels of university environments, and outlines policy initiatives that are already being introduced europe-wide to counter this problem.[16]

Challenges for younger women

When the professors gave advice from their own experience to the many young women aspiring to follow an academic career to professorship, the key advice was that young women should find ways to strengthen and develop themselves as human beings, to really find out who they are, and what their aspirations are in depth. From the majority of professors' experience, working life is not easy and the self-reflective work to understand your place in academia helps with the challenges of dealing with the competitive academic environment and can assist later with dealing obstacles that arise over time. Natalie Banerji talked about how, because she is really sure that being a professor is the career for her, she can face with strength the ongoing assessments, critical feedback, and heavy workload that comes with the job. Rachel Grange confirmed the importance of developing inner strength and self knowledge when she says that having mentors and good supervisors does help, but the key ingredient is growing your own self-belief, establishing your body of research, and finding ways to take the next steps yourself.

Eleni Chatzi wanted to remind female researchers to believe in their abilities, and to keep developing their self-confidence, because she sees how young women can be too self-doubting when, in reality, their achievements are equal to their male

[15] LERU Report, 2012, *Women, Research and Universities: Excellence without Gender Bias*. See also Prof. Macartan Humphries blog, *Gender discrimination in political science and the problem of poor allies*, http://www.macartan.nyc/comments/poor-allies/.

[16] The impact of unconscious bias and gender schemas was discussed in the research literature as early as 1999, in Virginia Valian's, *Why so Slow? The Advancement of Women*, MIT Press. The US National Science Foundation, established initiatives such as the ADVANCE projects, from 2001 focused on making changes in academia to combat bias. The LERU report, *Implicit bias in academia: a challenge to the meritocratic principle and to women's careers – and what to do about it*", was published in January 2018.

colleagues, and sometimes even better. All the women professors dealt with finding their way forward differently, but during the moments of uncertainty, they found ways to deal with failure and rejections. Their key advice is to give the negative experiences perspective and to learn from rejections and not to give up. When they experienced failure they tried again and were often successful subsequently.

Some of the younger professors shared the fact that, despite their positive career progression and excellent experiences, they still have to deal with, and ignore, the discouraging comments from male colleagues following professional success (such as receiving a new prize, appointment or grant): "Oh you only got that because you are a woman." This statement is a persistent and troublesome comment that female academics deal with at every level of the career. The professors advise that it is best ignored, however discouraging. The suggestion that women are getting favorable treatment is not based on the facts. In fact, Paolo Picotti investigated the award process for the EU grant application and found no evidence of favoritism toward women academics in her field. Young female researchers need to recognize that they achieve their PhD, postdoc position, research grant or prize only if their achievements meet the highest standards. It is not possible to prevent these comments, but you can control how they affect you.[17]

With regard to serious problematic experiences, such as sexual harassment, Effy Vayena advised that any harassment needs to be dealt immediately by seeking help from trusted colleagues or mentors to approach the responsible people in the university trained to deal with these situations. Such experiences should never be tolerated. Many female professors pledged to help improve academic environments, if necessary, by speaking up if they witness any problematic, negative or inappropriate behavior against female students, indeed against all students, who do not have the same power as professors in a university environment.[18]

Further advice to young women came from the professors who became mothers whilst continuing their research careers. They urge young women not to limit their career and life possibilities once they become a parent. They should rather explore all options available, and know that it is possible to find a way to be both a mother and to have a successful career. Central to decision-making is to understand who you want to be as a parent, and then to find the options that are right for you and your partner. In that sense you should seek support and advice on all possible childcare options for your current career stage. It is not always easy, because there can be cultural judgements about your parenting choices, but when you are clear about what you want, you will find solutions that will make it possible to have a family and a career.

[17] In certain disciplines negative comments against women scientists can lead to widespread press coverage. A lecture by a physicist who declared that women physicists were being unjustly favored, received international attention. https://www.theguardian.com/science/2018/oct/01/physics-was-built-by-men-cern-scientist-alessandro-strumia-remark-sparks-fury.

[18] Schubert, R. (November 2018). *Treating other people respectfully: The ETH Zurich RESPECT campaign supports its community by raising awareness of problematic behaviors.* Optics & Photonics News.

Nowadays, in Switzerland, an increased availability in university kindergartens exists, but the competition for places is still high. There is additional financial support, from the Swiss National Science Foundation, for young academic parents who can receive grants to pay for childcare.[19] More flexible provision for different parental needs is growing, such as small grants for parents who need childcare support to attend conferences. Importantly, in a number of universities, schemes are being created to support PhD and postdoc mothers by employing a laboratory technician, or a PhD student, during the maternity leave period so that their research work can continue.[20] Stefanie Hellweg advises young mothers to investigate the options out there, talk to people and not to limit your vision of parenting. She believes that if you think forward with an open mind in academia, there will always be solutions to what appear to be insurmountable obstacles; for example, it is possible to think of going on short research periods abroad, because they can be an adventure for the whole family. Indeed, these periods can add mobility experience to your CV. The professors who are mothers showed a range of approaches to managing a family and a career. Choose what works for you and your family.

Recommendations to aspiring professors from these conversations

During the meetings many of the professors gave thoughts on what might help young career researchers in finding a way forward in the academic career. A professorship is a highly demanding job requiring the ability to carry out unique research, to teach, to appoint and lead a group, to budget, to apply for grant applications to finance your group and (potentially) laboratories, to participate in departmental and university requirements and responsibilities, while developing your career through the stages of assistant professor, through the tenure process and then promotion to full professorship. It takes a lifetime because, once established you need to maintain this process, shoulder leadership responsibilities, and develop and consolidate your research field.

On the next page is a list of aspirational comments gathered from across the interviews on what is essential for a professorial career.

[19] The Flexibility Grant gives PhD and postdoc parents, with salaries paid by the SNSF, 1000 CHF per month toward the cost of childcare so that they can continue their careers.

[20] Department of Physics, ETH Zürich, introduced a scheme in 2018 of fellowships for postdoc mothers to finance a PhD student to support the early years of motherhood see Keller, U. & Garry, A. (December 2016). *Retaining postdoc mothers in an academic career.* Optics & Photonics News. The University of Basel has similar schemes for women researchers, "Get on Track" and "Stay on Track" initiatives. https://www.unibas.ch/en/University/Administration-Services/The-President-s-Office/Institutional-Development/Diversity.html.

- Make efforts to find and believe in your own career path.
- Discover your own research question and follow your curiosity.
- Think forward early: apply for grants and try to anticipate the next stages.
- Think broadly about possibilities, take unexpected opportunities, you never know where they might lead.
- Always keep learning about yourself, what you need, how you can be the best person you can be as a leader in the academic environment.
- Mentors, role models, and networks are of enormous support and can help you on the steps to independence: but you have to take the steps yourself.
- Following your own aspirations, talents, and interests will get you to a strong place in life, even if, in the end, academia is not the place for you.
- Understand what an academic career requires in depth.
- It is possible to develop over time the multiple skills needed to be a professor; you can work on the aspects that do not, initially, seem to be your strength.
- Look for the new positions yourself, don't wait for jobs to appear, or people to invite you, make contacts regularly with people in your field.
- Sometimes opportunities come when you are least expecting them and you need to be open to possibilities from chance meetings.

Several of these suggestions are about learning and developing central strength. Indeed certain professors emphasized the benefits of reaching out to get the formal help from coaches or psychologists, in order to get a deeper understanding of your inner strengths and motivations, the factors that might be blocking achievement and ambition, or could be fueling reactions to challenging experiences. Younger professors have attended or signed up for formal training courses on leadership skills and developing the techniques needed to be a well-rounded principle investigator or research group leader. There is a growing movement in universities to provide, and attend, the training that widens the range of professorial skills.

A professorship is a challenging and time-consuming job, but if you choose it as a career, and succeed in finding a permanent position, you will have incomparable independence in your working life and a career that many of the women professors in this book have described as "the best job in the world."

Conclusion

The book project began with the twin goals: to make visible exceptional female academics and, at the same time, to learn from them about diverse ways to find career success. The conversations here have illustrated a wide range of career experiences, journeys, and insights. What I had not anticipated in the early stages, was being introduced to the vast scope of cutting-edge research fields, where these women contribute new insights, innovations, and policies to academia and to society. Over time I heard about research being conducted in specialisms such as multiferroics, attosecond science, bioethics, the interactive geometry behind computer graphics, research on wound repair, cloud formation and its impact on climate, epigenetics, fungal immunology, proteomics, and new work in the psychology of teaching and learning science. These are just a few of the fields encountered in this book, and I was privileged to speak with all these women professors who are pushing the boundaries of knowledge across a multitude of disciplines.

Listening to each different professor's life experience and witnessing the diversity of origins, cultures, and aspirations, I began to understand that underlying all these differences was an unstoppable drive and commitment to curiosity about their topic, their question(s) and their chosen fields. This is what unites the women academics in this book, and took them through the various turning points, and decisions taken to try for the next academic step, whilst dealing with moments of doubt and self-questioning. In the end, they made a full-hearted commitment to their field, even if they were uncertain where it would lead and whether they would achieve a professorship. They found the way of life which is authentic to their happiness, dared to commit to what is involved, even when some were unsure they had the skills it took for such a complex career, or struggled against barriers that felt insurmountable. It was the research questions, and the love of addressing and solving intellectual challenges, the excitement of making a difference, and discovering something new, that pushed them onwards.

As we talked, the professors also brought into the light the people who had supported them, influenced them, and inspired them on their journey. There were generous supervisors, influential mentors, the senior people who knew how to step back and give space to their independent growth as academics, the communities who recognized the expansion of their expertise. Then, behind the scenes, there are the partners who made joint lives with them, kept open to all the possibilities that allowed them to grow as academics and people. These partners worked to find ways to make a dual career successful, making compromises, and also remaining open to different directions. The professors were underpinned by relationships that enabled them to grow, and those who had children, coparented with their partners. The majority of these supportive people are men, who worked with their partners, or within their academic communities to enable the growth and success of such talented women. It was very moving to hear this successful aspect of the professors' career stories, particularly as I see the current generation of young academics striving already to make dual careers work and, as new parents find ways to coparent, so can they move forward together with fulfilling lives.

I did raise the question of whether the women professors had experienced problematic situations because of their gender, and the majority focused on the positives in their career progression, and how they found ways to overcome career challenges in the transition periods, without relating them to a gender dimension. In advising young women, some raised the detrimental impact of unfriendly academic environments, whilst others mentioned the potential consequences of underlying and unquestioned behaviors, which could unconsciously have an unfavorable impact on the progress and promotion of women.[1] Associations of female academics, such as the ETH Women Professors Forum, have grown in number in the last decade, enabling the sharing of experience and advice between female professors and working to increase influence.[2] Indeed, the creation of the ETH Women Professors Forum underpinned the origins of this book. In addition, websites such as Academia.Net provide an archive of information on distinguished women professors in Europe, acting as a resource, and giving visibility to the large number of female experts active in universities on this continent.[3]

There is still not enough known in the public arena about the contribution of exceptional women professors to society, indeed a female journalist told me recently that her experience indicates that, from a public perspective, professors are still assumed to be men. I have had the privilege to speak with, and write about, twenty-three extraordinary women whose research expertise shone through, whilst they also spoke generously and openly of their journeys through a complex and fulfilling career, in a way designed to assist younger people considering a career in academia.

The female professors in this book have shown that intellectual curiosity, excellence in research, and individual dedication is a central aspect of their success; whilst illustrating also the power of positive and welcoming academic environments, with generous supervisors, supportive research communities, inclusive departments, and inspiring leaders. Those policy makers seeking to increase the number of female professors in academia, and at the leadership level, will find that these professors have provided information on the key ingredients to help achieve these important goals.

[1] Recent books on this topic can be found in the list of references at the end of the book.

[2] The Women Professors Forum website www.eth-wpf.ch provides profiles of female professors working across the faculties at ETH Zürich and EPFL Lausanne. Macfarlane B. and Burg D. Women Professors as Intellectual Leaders, 2018 has a list of a growing number of associations for female academics in the United Kingdom.

[3] Academia Net is a portal for excellent women in science. You need to be nominated to appear on this website http://www.academia-net.org.

Appendix 1: A list of general questions and topics adapted for individual interviews

When did you become interested in studying, or your topic?
What were the influential factors in choosing your university degree?
What was your experience of studying *X* in your *University of Choice*? Why did you choose there?
Were there many women in your classes? Impact of this.
PhD studies: why did you decide to study for a PhD?
You moved to *X* for your PhD studies? Why did you move? How was that experience?
Postdoc experience (s): what lead you to that group/field?
When did you decide/realize that you wanted to become a professor?
What steps did you take to achieve this?
How, and when, did you find your particular research expertise?
Who or what made a difference in your life choices?
How did you become a professor in your current position? What choices lead you here?
Dual career issues—was this an ingredient in the career?
What is your experience of being a professor and having children?
What were the positive experiences in your career?
What were the challenges at different stages in your working life?
Did you experience any negative impacts from being a woman in academia?
What advice would you give to young women from your experience?
What factors, or people, have supported your career?
What advice would you give to young career researchers interested in an academic career?

Glossary of terms

The A Level (Advanced Level) A subject-based qualification conferred as a part of the school leaving qualification offered by educational bodies in the United Kingdom. Obtaining A Levels, or equivalent qualifications, is generally required for university entrance, with universities granting conditional offers based on grades achieved.

Abitur A qualification granted by university-preparatory schools in Germany, which is conferred on students who pass their final exams at the end of their secondary education, usually after 12 or 13 years of schooling. It is the equivalent of a Matura or the International Baccalaureate.

Assistant Professorships Usually appointed to young academics who are 35 years or under at the time of appointment. Assistant professorships come with or without tenure-track and have an initial contract of 4 years.

Associate Professorships Appointed to academics who have achieved tenure in their careers.

Diplom Before the introduction of the bachelor's and master's degrees in Germany and Switzerland the standard science, engineering or business degree was the Diplom and could be, in several variations, obtained at several types of institutes of higher education, it could take between 4 and 6 years depending on the discipline. It has been replaced by bachelor's and master's courses.

ERC (European Research Council) Advanced Grant Applicants are expected to be active researchers who have a track record of significant research achievements in the last 10 years. The principal investigators should be exceptional leaders in terms of originality and significance of their research contributions. No specific eligibility criteria with respect to the academic requirements are foreseen.

ERC Consolidator Grant For researchers of any nationality with 7–12 years of experience since completion of PhD, and with same conditions above, though the funding can be 2 million euro over 5 years with extra funds for a start-up if the researcher moves countries within the EU.

ERC Starting Grant Researchers of any nationality with 2–7 years of experience since completion of PhD and with a scientific track record showing great promise and an excellent research proposal, can apply for this award. Research must be conducted in a public or private research organization based in a European Union member state or associated country. This institution has to offer appropriate conditions for the principal investigator independently to direct the research and manage their funding for the duration of the project. At ETH successful researchers become assistant professors for the fixed-term of the grant, which is a period of 5 years and receive funding of 1.5 million euros.

ERC Proof-of-Concept Grant Funding made available only to those who already have an ERC award to establish proof of concept of an idea that was generated in the course of their ERC-funded projects.

Full Professorship Achieved through a promotion process between 2 and 6 years after an academic is awarded tenure.

Gordon Research Conferences A group of prestigious international scientific conferences organized by a non-profit organization of the same name. The conference topics cover frontier research in the biological, chemical, and physical sciences, and their related technologies.

Gymnasium/Gymnase/Liceo An upper secondary school in Switzerland which prepares students for a Matura needed to enter university. You enter this school after taking a competitive exam at ages 12 or 15 years old. 20%–30% of Swiss students study at Gymnasium and go to university, whilst the other two-thirds follow the vocational dual-track training system, studying at school, whilst working as an apprentice in a company.

Habilitation The qualification is given to conduct self-contained university teaching and is the key for access to a professorship in many European countries. Despite all changes implemented in the European higher education systems during the Bologna Process, it is the highest qualification level issued through the process of a university examination and remains a core concept of scientific careers in many European countries. The degree is conferred for a habilitation thesis based on independent scholarship, which was reviewed by and successfully defended before an academic committee in a process similar to that of a doctoral dissertation. In some countries, a habilitation degree is a required formal qualification to independently teach and examine a designated subject at the university level.

Matura A Latin name for the secondary school exit exam or "maturity diploma" in Switzerland. The Matura is required for Swiss students to study at a university or a federal institute of technology irrespective of their subject choice (except for medicine, where the number of students is restricted).

National Centres for Competence in Research (NCCR) NCCRs program seeks to promote research projects of the highest quality with a particular emphasis on interdisciplinary approaches, but also on new, innovative approaches within the disciplines. It is, potentially, a 12-year funding instrument.

Oberassistant or Senior Scientist A scientist or project manager with extensive management duties who has supervisor/supervisory position working within a professorial group.

Swiss National Science Foundation (SNSF) It supports scientific research in all academic disciplines, from history to medicine and the engineering sciences. At the end of 2017, the SNSF was funding 5800 projects involving 16,000 researchers, which makes it the leading Swiss institution for promoting scientific research.

SNSF Professorship (renamed SNSF Eccellenza Professorial Fellowships in 2017) Aimed at outstanding researchers in all disciplines who have a doctorate or equivalent qualification and are pursuing an academic career, but who have not yet obtained an assistant professorship. Researchers can request their own salary at the local rates applicable to assistant professorships as well as project funds of up to 1,000,000 Swiss francs for 5 years.

Vordiploma A foundation course for a specialized Diploma course. It is, however, not an independent qualification, rather a stepping stone to a degree/Diploma course.

Bibliography

Women and gender equality

Bohnet, I. (2016). *What Works: gender equality by design.* Cambridge, Massachusetts: The Belknap Press of Harvard University.

Cornick, V. (2009). *Women in Science, Then and Now* (25th Anniversary Edition). New York: The Feminist Press.

National Academy Sciences, National Academy of Engineering and Institute of Medicine of the National Academies. (2005). *Beyond Bias and Barriers, fulfilling the potential of women in academic science and engineering*. National Academy of Sciences.

McFarlane, B., & Burg, D. (2018). *Women Professors as Intellectual Leaders.* London: Leadership Foundation for Higher Education.

Rosser, S. V. (2000). *Women, Science and Society: the Crucial Union,* Athene Series, Teachers College Press.

Sandberg, S. (2013). *Lean-In, women work and the will to lead.* WH Allen.

Valian, V. (1999). *Why so Slow? The advancement of women.* Cambridge Massachusett: MIT Press.

Warrior, J. (1997). *Cracking it! Helping Women to Succeed in Science, Engineering & Technology.* The Engineering Council.

Writing about women academics

Bostock, J. (2014). *The Meaning of Success: Insights from Women at Cambridge.* University of Cambridge Press.

Bertsch McGrayne, S. (1998). *Nobel Prize Women in Science, their lives, struggles, and momentous discoveries.* Washington: Joseph Henry Press.

Byers, N., & William, G. (Eds.), (2010). *Out of the Shadows, Contributions of Twentieth-Century Women to Physics.* Cambridge University Press.

Connelly, R., & Ghodsee, K. (2011). *Professor Mommy, finding work-finding balance in academia.* Rowan and Littlefield Publishers.

Evans, E., & Grant, C. (2009). *Mama PhD, women write about motherhood and academic life.* Rutgers University Press.

Madsen, L. D. (2016). *Successful Women Ceramic and Glass Scientists, 100 Inspirational Profiles.* New Jersey: Wiley.

Monosson, E. (2008). *Motherhood, the elephant in the laboratory, women scientists speak out.* Cornell University.

Pollack, E. (2015). *The Only Woman in the Room: why science is still a boy's club.* Boston: Beacon Press.

Rayner-Canham, M. F., & Rayner-Canham, G. W. (1998). *Women in Chemistry, their changing roles from alchemical times to the mid-twentieth century*. American Chemical Society and Chemical Heritage Foundation.

Robinson, J. (2010). *Bluestockings, the remarkable story of the first women to fight for an education*. Penguin.

Ruddick, S., & Daniels, P. (1977). *Working it Out, 23 women writers, artists, scientists and scholars talk about their lives and work*. Pantheon Books.

Wasserman, E. (2002). *The Door in the Dream, conversations with eminent women in science*. Washington: Joseph Henry Press.

Wertheim, M. (1997). *Pythagoras' Trousers, god, physics and the gender wars*. W.W. Norton & Company.

Women's lives

Barsh, J., & Cranston, S. (2011). *How Remarkable Women Lead, the breakthrough model for work and life*. Mckinsey & Company.

Bateson, M. C. (1989). *Composing a Life*. New York: Grove Press.

Beard M. (2017). *Women and Power*. Profile Books Ltd.

Fels, A. (2004). *Necessary Dreams, ambition in women's changing lives*. US: Anchor Books.

Academic research on narrating a career

Bowles, H. R. (2012). Claiming authority: How women explain their ascent to top business leadership positions. *Research in Organizational Behavior*, *32*, 189–212.

LaPointe, K. (2010). Narrating career, positioning identity: Career identity as a narrative practice. *Journal of Vocational Behavior*, *77*, 1–9.

Vinkenberger, C. J. (2015). Promoting new norms and true flexibility: Sustainability in combining career and care, *in Handbook of Research on Sustainable Careers*. London: Edgar Elder.

Index